T0335952

Barcodes for Mobile Devices

From inventory management in stores to automotive part tracking in assembly plants, barcodes are one of the most prevalent automatic identification and data capture technologies.

This book provides a complete introduction to barcodes for mobile devices where data captured in the device's camera can be interpreted by signal processing algorithms. The most relevant and up-to-date information, previously unavailable elsewhere or difficult to obtain, is presented. The focus throughout is on recent developments and two-dimensional (2D) barcodes, including the research and development steps towards colour barcodes for mobile devices, helping readers to develop their own barcodes. The authors also provide design details for their own novel colour 2D barcode, the Mobile Multi-Colour Composite (MMCC™) barcode, plus a coverage of RFID technology and one-dimensional barcodes.

This book is ideal for professional developers of barcodes for mobile devices who need the latest technical details and information on how to develop barcodes. It is also a useful reference for graduate students researching the field of barcode technology and mobile computing.

Hiroko Kato recently completed her Ph.D. in the School of Computer and Information Science at Edith Cowan University, Australia, where her research involved developing a novel 2D barcode. Her current research interests are computer vision and human computer interaction.

Keng T. Tan is the founder of GO-CDMA Limited, a private R&D company in Hong Kong. He is a technical expert in mobile computing and wireless communications. Dr Tan is also a registered IP attorney. In 1995 he won the Institute of Engineers Australia Medal.

Douglas Chai is a Senior Lecturer in the School of Engineering at Edith Cowan University. His current research interests include barcode technology, image processing, video coding and pattern recognition.

Barcodes for Mobile Devices

HIROKO KATO
Edith Cowan University, Australia

KENG T. TAN
GO-CDMA Limited, Hong Kong

DOUGLAS CHAI
Edith Cowan University, Australia

CAMBRIDGE
UNIVERSITY PRESS

CAMBRIDGE
UNIVERSITY PRESS

University Printing House, Cambridge CB2 8BS, United Kingdom

One Liberty Plaza, 20th Floor, New York, NY 10006, USA

477 Williamstown Road, Port Melbourne, VIC 3207, Australia

314-321, 3rd Floor, Plot 3, Splendor Forum, Jasola District Centre, New Delhi - 110025, India

103 Penang Road, #05-06/07, Visioncrest Commercial, Singapore 238467

Cambridge University Press is part of the University of Cambridge.

It furthers the University's mission by disseminating knowledge in the pursuit of
education, learning and research at the highest international levels of excellence.

www.cambridge.org
Information on this title: www.cambridge.org/9780521888394

© Cambridge University Press 2010

First published 2010

A catalogue record for this publication is available from the British Library

Library of Congress Cataloging in Publication data
Tan, Keng T.
    Barcodes for mobile devices / Keng Tiong Tan, Douglas Chai, Hiroko Kato.
      p. cm.
    Includes bibliographical references and index.
    ISBN 978-0-521-88839-4
    1. Bar coding–Equipment and supplies.   2. Cell phones–Equipment and supplies.   3. Pocket
computers–Equipment and supplies.   I. Tan, Keng T.   II. Chai, Douglas.   III. Kato, Hiroko,
1963–   IV. Title.
    HF5416.T36 2010
    621.3845´60687–dc22   2009054010

ISBN 978-0-521-88839-4 Hardback

Contents

The colour plate section is to be found between pp. 182 and 183.

Preface

This book was written for scientists, researchers and those who are interested in the science and development of barcode technology for mobile devices. It is a comprehensive work covering the history and evolution of this technology. The book also presents a substantial compilation of different barcodes used or designed for mobile devices and mobile applications. Nonetheless, this collection will never stay complete for long as this is a rapidly evolving field, with new mobile barcodes being invented on almost a yearly basis. In fact, within this book, we have presented our very own novel two-dimensional (2D) barcode, designed especially for mobile devices such as camera mobile phones.

Also presented herein are the techniques and development process for our novel 2D barcode, the Mobile Multi-Colour Composite (MMCC™). Hence, we envision that this book will be a useful reference for those who are interested in designing and developing barcodes for mobile devices.

Finally, we do hope that this book will be the first of many to come that will fill the current gap in the publications in this area of interest.

Acknowledgements

Any substantial work of substance would not be possible without the kind contributions and supports of others. For continuous support and encouragements from Cambridge University Press, we wish to thank especially Dr Phil Meyler and Ms Sabine Koch.

For the personal support of our loved ones, our respective acknowledgements are as follows.

Dr Hiroko Kato wishes to thank her husband, Hisanori, for his continued support and patience and her children, Masanori and Yukino, for their constant support and understanding and all the help they gave with household chores while she focused on completing this book.

Dr (Alfred) Keng T. Tan wishes to thank his wife, Jo, for her understanding, love and support and his two girls, Kylie and Andrea, for the joy they bring to our family.

Dr Douglas Chai wishes to thank his wife, June, for her love and support, his children, Ethan and Amber, for making life so much fun, and his mother, Helen, for her wisdom and kind patience.

1 Introduction

1.1 Overview of barcode technology

First of all, it is important to know the relationship between a 'code' and a 'symbol' in this context:

The shorthand used to represent the verbal description of an item is called the product identification code. The product identification code is shortened to the word 'code'. The use of the word 'code' should not be confused with the barcode, which is technically called the symbol. A barcode symbol is used to identify people and places as well as a product [1, p. 44].

1.1.1 Definition of barcode

The definition of a barcode is given in the International Organization for Standardization and the International Electrotechnical Commission (ISO/IEC 1976 2-2): Information technology – Automatic identification and data capture (AIDC) techniques – Harmonised vocabulary – Part 2: Optically readable media (ORM). Since it only includes one-dimensional barcodes, the definition given by the Japanese Standards Association (JSA) is preferred.

A barcode is a machine-readable representation of information that is formed by combinations of high and low reflectance regions of the surface of an object [2], which are converted to '1's and '0's. This definition includes both one-dimensional and two-dimensional barcodes. Originally, information was encoded into an array of adjacent bars and spaces of varying width and that is where the word 'barcode' is derived from. This type of barcode is called a linear one-dimensional (1D) barcode. The 1D barcode symbologies can be read by a scanner that sweeps a beam of light across the barcode symbol in a straight line [2].

By replacing the bars and spaces with dots and spaces arranged in an array or a matrix, the density of data within a given space can be increased. The resultant symbol is called a two-dimensional (2D) barcode. Unlike their 1D counterparts, the 2D barcode symbologies need a scanning device that is capable of simultaneous reading in two dimensions, i.e. vertically as well as horizontally. Strictly speaking, most 2D symbologies do not use bars for encoding data, and '2D code' may be more appropriate for symbologies of this class. However, in general both 1D and 2D barcodes are included as different forms of barcode technology, distinctive from other automatic identification technologies such as *radio frequency identification* (RFID). Hence, in this book, the term two-dimensional barcode (2D barcode) is adopted to describe symbologies that belong to this class of barcode technologies.

In addition to 1D and 2D barcodes, three-dimensional (3D) barcodes exist. A 3D barcode, also called a 'bumpy barcode', is any linear (1D) or 2D barcode that is directly embossed on the surface of a material [3]. Three-dimensional barcodes use spatial elements, or a bumpy aspect of symbols, to generate high and low reflectance instead of visual contrast between different colours. Since this type of barcode cannot be handled by mobile devices, 3D barcode technology is out of the scope of this book. Thus, it will be touched upon only when it helps to explain the entire barcode technology in relation to 1D and 2D barcode technologies.

The terms barcode symbology, barcode symbol and barcode label should not be confused with one another. A *barcode symbology* is a scheme or a set of rules for encoding information into a barcode format. A *barcode symbol* is a graphical representation of the symbology which is composed of symbol elements such as adjacent bars and spaces or dots and spaces. For example, a 1D barcode symbol includes a leading and a trailing quiet zone, a start and a stop character, data characters and checksum when required. A *barcode label* is a label that carries a barcode.

Barcode is also called 'bar code'. Throughout this book, however, the term barcode will be used.

1.1.2   General anatomy of 1D and 2D barcodes

Numerous barcode symbologies have been developed and each has its distinctive features, as described in later chapters. However, there are common structural features shared by these barcodes. It is appropriate to present the anatomy of each type of barcode, namely, 1D and 2D barcodes, so as to provide a picture of each barcode technology.

Anatomy of 1D barcodes

In general, a 1D barcode symbol is made up of a start character, data characters, a stop character and quiet zones before the start character and after the stop character. Figure 1.1 presents the data structure of Code 128.

The following are brief descriptions of each component.

Quiet zone

The quiet zone is the area of high reflectance allocated before the start character and after the stop character of a 1D barcode symbol. The quiet zone is necessary for most barcode symbols to be reliably read and is included as part of the symbol. However, newly developed barcodes such as those of the GS1 DataBar family do not require quiet zones.

Start and stop characters

The start character is unique and is normally located at the leftmost edge of a horizontally oriented symbol. The stop character is also a unique character, located at the rightmost edge of a horizontally oriented symbol. These characters provide a scanner with reading

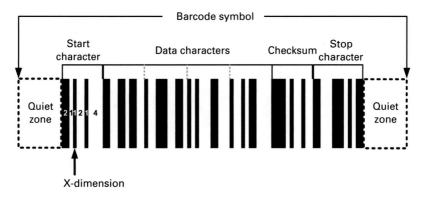

Fig. 1.1   General anatomy of 1D barcode: example – Code 128 symbol.

instructions such as the scanning direction and when to start or stop reading. The start and stop characters also allow barcode symbols to be read bi-directionally.

Checksum
Barcode symbols include a mandatory or optional checksum (also called the check character or check digit), whose value is used for the purpose of performing a mathematical check to ensure that the barcode is accurately decoded.

Data characters
The data characters or message characters appear after the start character. In Figure 1.1, four characters, !, 1, A, a, are encoded from left to right.

X dimension
The X dimension or module width is the dimension of the narrowest element (either bar or space) of a barcode symbol. The wider elements of the symbol are generally referred to as multiples of the X dimension. It is usually stated in mils, or one thousandths of an inch. In general, the larger the X dimension is, the more robust barcode reading can be achieved. However, a larger X dimension increases the size of a barcode symbol and, as a result, requires a wider space for printing. Most symbologies require that the quiet zone be 10 times the X dimension of the symbol, or one quarter of an inch, whichever is greater.

The data encoded in 1D barcode are vertically redundant. This enhances the reading robustness, allowing symbols to be read correctly even when they are partially damaged. The higher the bar heights, the greater the probability that at least one path along the barcode will be readable [3]. However, the bar height increases the amount of space required for printing symbols. In contrast, two-dimensional barcodes encode data not only vertically but also horizontally. This significantly increases the data capacity of 2D barcodes, yet they lack physical redundancy and thus have an inferior reading robustness.

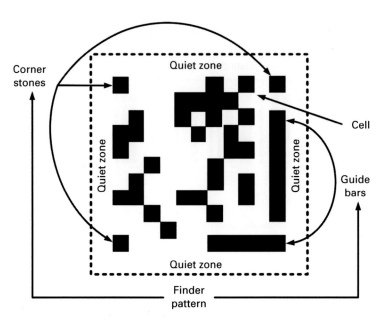

Fig. 1.2   General anatomy of 2D barcode: example – Visual Code symbol.

Anatomy of 2D barcodes

Currently a variety of 2D barcodes exist and some of them form distinctive shapes. However, most 2D barcodes, especially those used for mobile devices, are shaped as square arrays. With few exceptions, 2D barcodes contain common basic components: a finder pattern, a data area and a quiet zone surrounding the graphic symbol. Figure 1.2 presents the anatomy of an example 2D barcode called Visual Code.

Finder pattern

Each 2D barcode symbology has a unique finder pattern that is used to detect and locate the 2D barcode symbol. A finder pattern can be an L-shaped border, a square frame, a round cell located at the centre of the symbol, a plurality of dots arranged in a pre-determined geometric pattern and so forth. A finder pattern is also used to compute the properties of a symbol (e.g. size, location and orientation) and correct them when required. In Figure 1.2, square cells at the three corners of the symbol (i.e. the corner stones) and two bars (i.e. the guide bars) comprise a finder pattern. In addition to the detection of the symbol, the corner stones are used to correct symbol distortion. The guide bars help to detect and correct the orientation of the symbol. Most 2D barcodes have error detection and correction capability, which increases the robustness of their reading. With an error detection and correction technique, the original data can be accurately retrieved even when the symbol is partially damaged, except when the finder pattern or part thereof is damaged. In the latter instance the barcode may not be successfully decoded or the barcode reading time may be increased significantly even if the barcode is eventually decoded successfully.

Data area

Within the graphical symbol in Figure 1.2, the area other than that occupied by the finder pattern is the data area. It comprises dots and spaces, called cells or sometimes modules, arranged in an array. The cells can be square, circular, triangular, hexagonal or any other geometric shape.

Quiet zone

Most 2D barcodes require a quiet zone surrounding the graphical symbol. The quiet zone helps detection of the barcode symbol, clearly separating the symbol from its background.

Most 1D barcodes work as an index to a backend database, encoding only a limited amount of data whereas most 2D barcodes can operate as a portable database, owing to their much greater data capacity. However, as a whole all barcode technologies have common advantageous and disadvantageous features.

1.1.3 Advantages and disadvantages of barcode technology

The 1D, 2D and 3D barcode technologies operate differently. Each has its strengths and shortcomings. In general, however, all barcode technologies have the following advantages and disadvantages.

Advantages

The advantageous features of barcode technology include the following:

(i) Fast, accurate and reliable keyless data entry can be achieved.
(ii) The technology is versatile and operates inexpensively using paper and ink.
(iii) The technology provides real-time information and allows the decisions to be made accordingly for ongoing operations.
(iv) Printed tags can last a long time without any deterioration. In a harsh environment, direct marking can be used.
(v) The barcode technology can offer secure operation. Once data are encoded in the format of a barcode, the data cannot be changed without physical alteration.

Disadvantages

There are some disadvantages in using barcode technology. The disadvantages of barcode technology are as follows:

(i) To read barcode symbols, a clear line of sight is required.
(ii) More than one barcode symbol cannot be read at a time.
(iii) Barcode technology does not have the ability to scan an object inside a container or a case.
(iv) The reading distance is rather short. Some scanners need physical contact to read the barcode symbols.

Each barcode format, i.e. 1D barcode, 2D barcode and 3D barcode, has its own advantages and disadvantages. Furthermore, each barcode symbology in any barcode format has its own advantages and disadvantages. These will be discussed in detail in Chapters 2 and 3. In the next section, the current use of barcodes will be introduced.

1.1.4　Barcode applications

The barcode technology is versatile and myriads of barcodes are used ubiquitously (see Appendix A). New barcode applications are being developed and introduced constantly. Instead of listing all the barcode applications currently in use, this book will provide the reader with information on where and how barcode technologies are used. The focus of the book is on the novel and innovative use of barcodes for mobile devices. Such applications are introduced in this chapter and discussed in greater detail in Chapter 4.

Point-of-sale systems in the retail, wholesale and grocery industries

The typical and most well-known barcode users have been retailers in the supply chain. At the point of sale (POS), they are the conduits as well as the key to the information flow [4]. In fact, retailers took the initiative in introducing barcode technology to the world.

The 1D barcodes called the *European Article Number* (EAN) and the *Universal Product Code* (UPC) have been main players for item identification at the POS. With EAN or UPC, also known as the International Article Numbering (IAN) or World Product Code (WPC), any item can be uniquely identified throughout the world. Owing to their limitation in data capacity, however, EAN and UPC will be replaced with the GS1 DataBar family, which will be explained in detail in Chapter 2.

The EAN and UPC barcode systems at POS provide accurate real-time information on products. This significantly improves the management of products, enabling efficient stock, inventory and shipping control.

Logistics in warehouses

While the products themselves are labelled with EAN or UPC, the cartons of products are usually labelled with specific symbologies for physical distribution. Interleaved 2 of 5 (ITF), a symbology that encodes packaging levels as well as the trade item number, is the most common approach in this area.

In recent years, Electronic Data Interchange (EDI), the standardised electronic format for business transactions, has been globally used in the field of physical distribution. In EDI, a 14-digit *Global Trade Item Number* (GTIN) encoded in ITF-14 is normally used as an identifier to look up product information in a database.

Barcode technology enables the efficient logistics management, for which accurate information on products such as name (or ID), quantity, price and location at a given time are required. In warehouses, for example, barcodes help to simplify inventory searching and control, prevent shipping mistakes, enable first-in first-out execution based on expiration-date control and improve traceability by means of manufacturing history control.

Manufacturing and industry

In the industrial sector, barcode symbologies that have the ability to encode extra data other than keys to a backend database have been preferred. Hence the early trend was to adopt 1D barcodes that are capable of encoding alphanumeric data such as Code 39 and Code 128. Since the invention of 2D barcodes, the trend has shifted to 2D symbologies, owing to their superiority in data capacity and efficiency in space.

Code 39 is mandated for some automotive industries and Department of Defense labels in the USA. Two-dimensional symbologies such as Data Matrix, QR Code, PDF417 and MaxiCode have also been used commonly in the automotive industry, owing to their greater data capacity. The Automotive Industry Action Group (AIAG) took the initiative in promoting 2D barcode technology throughout the industry. The AIAG developed specific application guidelines for specific 2D barcode use. It recommended Data Matrix for part marking and tracking, MaxiCode for freight sorting and tracking and PDF417 for general applications such as quality conveyance, production evidence, production broadcast, configuration management, material safety data sheets, shipping and EDI.

In manufacturing, 2D barcode technology has been adopted for two main reasons: part identification and product assembly process control. With 2D barcodes, each item can be accurately identified and tracked without accessing a database. Moving data files affixed to the products also enables the system to track an individual product throughout the assembly process and, as a result, to pinpoint the location of a problem and fix it. Such systems can save time and cost for product assembly, resulting in much greater productivity.

The electronic industry is also one of the pioneering users of 2D barcode technology. Such 2D barcodes as the Data Matrix, which can store data within a limited space, have been the best choice for electronic assemblies since free space on each item is scarce. In fact, the Electronic Industries Alliance (EIA) has specified Data Matrix in its component marking standard (EIA-706) and product marking standard (EIA-802). The former standard covers the use of 2D barcodes for marking electronic components of first-level assemblies, whereas the latter standard covers both labels and the direct marking of products, including the testing procedures for label-adhesive characteristics and mark durability. Two-dimensional barcodes can be used for component tracing, component tracking, automated manufacturing and process control, inventory and configuration management, automated inspection, quality control and testing, product marking and so forth.

Healthcare industry

The healthcare industry also showed an early interest in barcode technology. The importance of accurate data entry and identification is tremendous in the healthcare industry, where people's lives are at stake. A minor mistake could result in a tragic malpractice. Nonetheless, medical institutions are flooded with many different kinds of medicine and a variety of medical equipment. The expiration dates are fixed, for most of them, for safe and efficient use. Once they have expired they must not be administered or used no matter how costly they are. As a safety measure, efficient replenishment is necessary.

To ensure safe treatment and its continuous improvement, the establishment of traceability systems is very important. The data on a patient's treatment, for example, which medicines and/or equipment have been used, when and where, by whom and so forth, need to be precisely recorded. This can help to minimise escalation of the damage even when malpractice occurs. In response to these demands, barcodes can be used for:

 (i) error and resultant malpractice minimisation;
 (ii) inventory control and replenishment;
(iii) traceability systems; and
(iv) fast and accurate billing.

In the USA, since 1987 the Health Industry Business Communications Council (HIBCC) has administered the Health Industry Number (HIN) system to eliminate expensive and inefficient administrative cross-referencing tasks. The HIN, a randomly assigned nine-character alphanumeric identifier, is designed for identification of all trading partners when communicating with each other via computer [5]. It can identify not only specific healthcare facilities but also specific locations and/or departments within them [5].

For product identification, in 1995 the United States Department of Defense (DoD) mandated the use of the Universal Product Number (UPN) for the primary product identification of medical and surgical products [6]. Each product at all levels of packaging was to be assigned a unique UPN that can be created using either the Code 39 format standardised by HIBCC or the global standard format, GS1-128 (formerly known as UCC/EAN128). The UPN is beneficial for all interest groups, ranging from customers and manufacturers to distributors, since it facilitates the use of EDI, which, in turn, enhances simple, fast and accurate ordering and distribution.

The Food and Drug Administration (FDA) in the USA has managed and controlled drugs and medical devices using the National Drug Code (NDC) and the National Health Related Item Code (NHRIC). The former is used for identifying pharmaceutical products while the latter identifies medical and surgical devices. Both codes can be directly incorporated into the Global Trade Identification Number (GTIN) [6].

The movement for unambiguous identification of medical and surgical products throughout the supply chain also took place in Europe. In 1995, the European Health Industry Business Communications Council (EHIBCC) adopted the UCC/EAN128 (now known as GS1-128) format for labelling medical and surgical products, which has become the de facto global standard.

When very small items that cannot include linear barcode formats (e.g. unit-dose packaging) are to be labelled, space-effective, high-density, GS1 DataBar symbologies and 2D barcode symbologies such as MicroPDF417 and Data Matrix can be used instead.

Publishers
International Standard Book Numbering
The *International Standard Book Number* (ISBN) developed in ISO 2108/1972 is globally used for the numbering of published books. An ISBN, which used to be a unique 10-digit number, is assigned to each edition and variation of a book.

Table 1.1. The structure of the ISBN

Component	Number of digits	Assignor	Note
Prefix	3	GS1	978 or 979
Group identifier	up to 5	International ISBN Agency	A group of countries sharing a language
Publisher code	up to 7	National ISBN Agency	
Item number	up to 6	Publisher	
Checksum	1		Modulo 10 with $1\times$ and $3\times$ weighting on alternate digits

The International Article Numbering Association Number (EAN)[1] and the International ISBN Agency have reached agreement on the coordination of the EAN and ISBN systems. This coordination allows the EAN symbols to be used for representing ISBN numbers, by addition of the prefix digits 978 to the ISBN [4]. The prefix is assigned for the exclusive use of book coding throughout the world.

Since 1 January 2007, a revision to the ISO standard governing ISBNs has come into effect, which rules that unique 13-digit numbers shall be assigned for ISBNs. This is needed because there is a shortage of numbers in certain ISBN categories. It has led the ISBN to consist of 13 digits if assigned after 1 January 2007 and 10 digits if assigned before that. The former is called ISBN-13 and the latter ISBN-10. The ISBN-13 has the same format as the ISBN-10 when represented in the form of an EAN symbol. Table 1.1 presents the structure of ISBN-13. Removing the prefix from ISBN-13 numbers gives the structure of ISBN-10. The prefix 979 has been assigned by the ISBN agency as a future extension.

International Standard Serial Numbering

For a number of serial publications such as magazines, the *International Standard Serial Number* (ISSN) is used worldwide. The ISSN system was developed on the basis of ISO 3297/1975. The International Article Numbering Association EAN and the International Centre for the Registration of Serial Publications agreed on the coordination of EAN and ISSN systems [4]. The prefix 977 has been assigned to the ISSN agency for its exclusive use of coding periodicals and journals throughout the world.

Innovative and foreseeable barcode use

In recent years, the combination of two mobile technologies, namely, 2D barcodes and camera phones, is gaining popularity as a promising ubiquitous computing tool. With the built-in cameras, mobile phones can work as scanners, barcode readers and portable data storages while maintaining network connectivity. When used together with such camera phones, 2D barcodes work as a tag to connect the physical world and the digital world.

---

[1] EAN originally stood for European Article Number but, since its adoption as an international standard, it is now also used as an abbreviation for the International Article Numbering Association Number.

The most popular application is the linking of camera phones to Web pages via 2D barcodes. Such applications allow users to have 'always on' access to on-line activities and information such as on-line shopping and real-time information about public transport schedules and events. Saved on mobile phones, 2D barcodes can also be used as portable data files such as e-tickets or e-coupons. They can be purchased and exchanged via the Internet. E-tickets shown on the phone display can be scanned and verified at the check-in counter or reception with no attendants, which results in faster ticket handling. Furthermore, no paper is used, making the procedure environmentally friendly.

Considering the rapid improvements in reading distance as well as decoding reliability, availability and functionalities, applications that will allow camera phones to interact with 2D barcodes on a plasma display will no longer be a Utopian dream. With 2D barcodes, users can access the information they need at any time, anywhere, regardless of network connectivity. Numerous 2D barcodes have been developed to implement innovative applications, and this trend is continuing.

In Chapter 3, we describe 2D barcodes in greater detail; this is followed in Chapter 4 by a discussion on how barcode applications have evolved.

1.2   Organisation of the book

Since its first emergence, barcode technology has evolved significantly. Different types of barcode technologies have been developed and there is a variety of different barcode symbologies in each format. Each barcode symbology was developed to optimise its use for certain applications. In the first few chapters (i.e. Chapters 2–4), this book provides a comprehensive description of barcode technologies and some well-known barcode symbologies in each barcode format together with their application, with an emphasis on the 2D barcodes used for mobile applications. The latter chapters are dedicated to explaining the concept of barcode operation, followed by the development of a novel colour 2D barcode. They include an explanation of the technologies that improve the robustness of barcode symbols and their encoding and decoding algorithms (in Chapter 5), a novel colour 2D barcode development (in Chapter 6) and evaluation of the developed colour 2D barcode (in Chapter 7).

2 Barcode technology evolution

Since the first barcode system was first introduced in a Cincinnati supermarket in the United States of America (USA) 40 years ago, this technology has co-evolved together with printing and scanning techniques. The printed tags are inexpensively produced as the costs of papers and printing devices have never had a dramatic increase. This has fostered the inexpensive operation of barcode systems and, in turn, their widespread adoption. The scanning technology has also had a great impact on the development of barcode technology, as without effective scanning barcode systems cannot operate. In fact, the lack of feasible scanning systems delayed the development of early barcode systems.

Different types of barcode technology, namely, one-dimensional (1D), two-dimensional (2D) and three-dimensional (3D), have been developed to meet user needs, along with numerous barcode symbologies. Each barcode symbology has unique features to meet specific requirements. As a result, a wide variety of barcode technologies is currently available, which not only meet user needs but have become a necessity in our daily lives.

In this chapter the history of barcode technology is introduced, and this is followed by a detailed explanation of how 2D and 3D barcodes have evolved since the birth of 1D barcodes (also called linear barcodes). Next, the barcode systems and symbologies that are commonly used at present are described. Finally, we forecast how each barcode technology will progress, taking into account how their interactiveness has promoted their advancement. Furthermore, the adoption of other emerging technologies such as radio frequency identification (RFID) is also predicted.

2.1 History of barcode technology

The concept of an automatic identification and data capture system has a relatively long history. According to [4] its origin dates back to the early 1800s, when devices were patented as reading aids for the blind. There are different types of automatic identification and data capture systems such as *optical character recognition* (OCR) systems, RFID systems and *magnetic stripe* technology as well as barcode technology (see Appendix B). Each has its own advantages and disadvantages.

One of the most distinctive advantages of barcode use is its inexpensive operation. LaMoreaux [4] claimed that this way of printing information in machine readable form is the least expensive. Rekimoto and Ayatsuka [7] support this idea, stating that 'Printed tags are probably the least expensive and most versatile tagging technology: they can be easily made by normal printers, attached to almost any physical object and recognised by mobile readers.' Another advantage is that it is the simplest way to enter information into a computer without keystrokes [4, p. 3].

Technological advancement especially in computing accelerated the search for better data entry methods. Data recognition patterns, similar to the current barcode technology, were developed in the 1950s or earlier [4]. In October 1949 a patent application entitled 'Classifying apparatus and method' was filed by two graduate students from Drexel Institute of Technology, Norman J. Woodland and Bernard Silver. The patent was issued on 7 October 1952 [9]. The symbol invented by Woodland and Silver is known as a bull's eye since it is composed of a series of concentric circles rather than straight lines (i.e. bars). Figure 2.1 shows the original bull's eye symbol.

This was the first step towards the present linear barcode. In fact, Woodland initially designed a barcode symbology made up of narrow and wide vertical lines but this was replaced with concentric circles to allow omnidirectional scanning [3]. However, the adoption of early barcode symbology was hindered by inadequate pattern reading or scanning technology. Despite the advances in computer design, there were few means to scan the patterns at the time. At the time the only devices used for reading patterns were transducers. The predominant types of scanner use pen and flashlight-type light sources and mirrors. Hence, the arrival of the moving-beam laser scanner in the 1960s, the subsequent invention of the charge coupled device (CCD) scanner and similar pixel readers was very important for the later development of barcode systems.

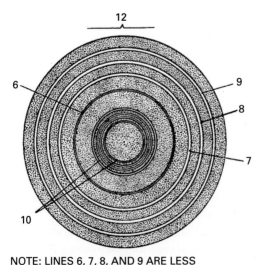

NOTE: LINES 6, 7, 8, AND 9 ARE LESS
REFLECTIVE THAN LINE 10

Fig. 2.1   Image of 'Bull's eye' symbol submitted to the United States Patent and Trademark Office (adapted from [9]).

An innovative barcode scanner system developed by David Collins and Chris Kapsambelis at Sylvania/GTE in 1962[1] called KarTrak is noteworthy. KarTrak was an optic scanning system that illuminated a barcode made up of horizontal bars of reflective red, white and blue tape on a non-reflective black background. The system operated by measuring the width of the horizontal bars on the labels appearing on the sides of railroad cars, using an optical sensor that measured each bar's reflected light [10]. This was the first industrial barcode application and more importantly, the first attempt to use colour elements in barcoding.

The system was developed in response to the requirement, presented in 1959 by a group of railroad research and development managers, to solve the problem of gathering owner and serial number information from moving railroad cars. In 1967 the railroads in North America adopted the tracking system. However, insufficient investment in training, maintenance and equipment led the system to be abandoned in 1974. Indeed, by 1975, up to 20% of the barcodes had become unreadable owing to inadequate maintenance [10].

In 1968 Collins co-founded Computer Identics Corporation, the first company to design and manufacture barcode scanning equipment. It invented Code 2 of 5 and then Interleaved 2 of 5. For scanning, the innovative fixed helium–neon (HeNe) laser was invented and used. This company's barcode system led in 1972 to the first successful use of barcoding on a large multi-plant scale for identifying engines and axles at General Motors in Lansing and Pontiac, Michigan [4]. The Interleaved 2 of 5 barcode format was used and this was quickly adopted by all automotive manufacturers [11].

The initial purpose of product coding in the USA in the 1960s and 1970s was to minimise human error during data entry and to process data efficiently. That is, barcode technology was to be used to improve the accuracy and speed of computer data entry. In fact, the substitution error rate[2] of 1D barcode is much less than an experienced manual operator's rate, which is about one substitution error per 300 characters entered [8]. The needs of retailers, wholesalers and grocery manufacturers played a vital role in the adoption and standardisation of barcodes in the USA. The use of barcode also enabled retail stores to register data on each product at the point of sale (POS) and obtain real-time information on products, which, in turn, allowed the efficient monitoring and control of products.

Some experiments and trials were conducted by supermarkets and retailers. For example, the company RCA installed one of the first scanning systems at a Kroger store in Cincinnati in 1967. It was aiming to observe the economic effect of the barcode-scanning system at the POS [12]. The product codes were represented by a bull's eye quite similar to the one developed by Woodland and Silver (see Figure 2.2). In 1972, Migros and Zellweger in Switzerland also carried out a collaborative experiment on an automatic point of sale system (APOSS) [12], which, apart from the central ring, used only the upper half of the bull's eye (see Figure 2.3). Both experiments proved the promising

---

[1] Sylvania Electronics merged with General Telephone to form General Telephone and Electronics (GTE) in 1959 and was named Sylvania Lighting under GTE ownership. With the acquisition of GTE's SYLVANIA lighting division by OSRAM GmbH in 1993, it has operated as OSRAM SYLVANIA Inc. in the USA.

[2] A symbology's estimated substitution error rate is a measure of the accuracy of the scanned data.

Fig. 2.2   Example of RCA/Kroger barcode symbol (adapted from [12]).

00110   00231   00320   00215   00160   00242

Fig. 2.3   Examples of the Littion–Zellweger/Migros barcode symbol (adapted from [12]).

outcome of barcode use. However, to make such systems work efficiently the industry needed to agree on a standard coding scheme open to all equipment manufacturers to use and adopted by all food producers and dealers [3].

In 1969, in the USA, the National Association of Food Chains (NAFC) commissioned Logicon Inc. to develop a proposal for an industry-wide barcode system. The resultant barcode, namely, the Universal Grocery Products Identification Code (UGPIC) was made public in the following year [3]. On the basis of the recommendations of the Logicon report the grocery industry formed an ad hoc committee, the Uniform Grocery Product Code Council (UGPCC) [4], which decided to separate the problem of designation of the code from that of the selection of a symbol.

In 1971 a 10-digit numeric code structure was selected, which was followed by the selection of a symbol on 30 March 1973 by a Symbol Selection Subcommittee that represented retailing, wholesaling and manufacturing. In April 1973 the grocery industry formally established the Universal Product Code (UPC) as the standard barcode symbology for product marking, and this is often considered to be the birth of the one-dimensional (1D) barcode [4]. With minor changes, the symbol set developed by George Laurer from IBM was selected as the UPC from eight candidates and is still in use in the USA today [13]. Figure 2.4 presents examples of the candidate code symbols.

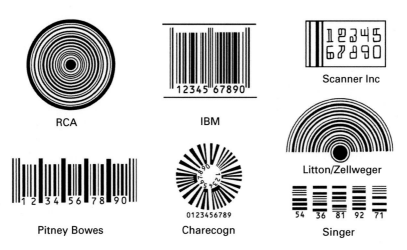

Fig. 2.4   Examples of proposed codes as UPC (adapted from [12]).

The UPC was the first linear barcode symbology widely adopted. As the name UPC indicates, the whole code and symbol selection has been conducted taking it for granted that all the interest groups (e.g. grocery manufacturers, wholesalers and retailers) use the same symbol structure, enabling each code to be uniquely identified. The entire system works in an efficient and effective way only when all groups involved use codes that have the same structure. This led to the formation of the committee and its collaborative work in the USA.

In the late 1970s, the Universal Product Code Council Inc. took over total control of the UPC administration from Distribution Code Inc., which had managed the system since adoption of the code [4]. The council manages the code in accordance with policies and procedures established by a 21-member Board of Governors, drawn from all sections of the council's membership (i.e. grocery retailers, wholesalers and manufacturers). Administrative matters handled by the council include:

 (i) assigning the five-digit UPC manufacturer identification number;
 (ii) providing UPC item assignment guidelines;
 (iii) providing an information and enquiry service to answer questions about UPC; and
 (iv) administrating council membership [4].

Foreign interest in UPC led to the establishment of the European Article Numbering Association and its adoption of the *European Article Numbering* (EAN) code format, similar to that of the UPC, in 1977 [13]. The main characteristics of EAN are that:

 (i) it can be omnidirectionally read;
 (ii) it is compatible with UPC;
 (iii) it does not require manufacturers' special printing and wrapping; and
 (iv) it can be decoded at a speed within 100 inches per second [12].

Whereas UPC code format was established considering only code use within the United States, the EAN format was adopted under consideration of the global use of the code. Hence, an EAN code includes two or three country identification numbers. More than one hundred countries all over the world have adopted a code that has the same structure as the EAN code and has the EAN format. Hence, EAN code has become the global standard 1D barcode. The use of a few globally standardised codes has brought enormous benefits to all the interest groups, manufacturers, wholesalers, retailers and consumers. It is the key factor behind the proliferation of 1D barcode.

In 1981, the United States Department of Defense (US DoD) adopted the use of barcode symbols called Code 39 for logistics applications of automated marking and reading symbols (LOMARS). Thus LOMARS is a special application of Code 39 for marking all products sold to the US military. The Code 39 symbology has become one of the most widely used industrial barcodes owing to its ability to encode alphanumeric data. The Automotive Industry Action Group (AIAG) and the Electronic Industries Alliance (EIA) mandated Code 39 for labelling in 1984 and in 1987, respectively.

The timeline of 1D barcode development is presented in Appendix C.

2.2   Development of barcode technology

Since the official introduction and enormous success of the use of 1D barcodes, especially in distribution systems, there has been a growing demand to encode more data than 1D barcodes can accommodate, especially in fields such as the automotive and electronic industries. Toyota Motor Corporation in collaboration with Denso Wave Inc., a Japanese automotive supplier, devised a barcode symbol that is similar to the stacked type of two-dimensional barcode currently used. The devised code consists of a stack of linear barcodes called Codabar: five-row Codabar, 11-row Codabar and again five-row Codabar.

The code was utilised in a parts-movement system called Kanban that enables efficient management of parts flow along a production line. Kanban, meaning a plate or a card with necessary information for production (in this context), contributes to productivity enhancement in Toyota's 'Just in Time' system. A project for a new system that utilises a Kanban with 2D-barcode-like symbologies began in 1978 [13]. This led to the development of two-dimensional barcodes that can encode 10 to 100 times more data than their 1D counterparts, which are only capable of encoding up to approximately 20 alphanumeric characters. Despite an increase in data capacity, in general 2D barcodes require less space than 1D barcodes when an identical set of data is encoded, owing to their matrix format.

Initially, 2D barcodes were considered to be advantageous in applications where space is a problem, where a database cannot be accessed or where high-speed sorting and/or processing is required [4]. Therefore, they were developed mainly for industries such as healthcare and electronics industries that need to handle very small products. However, the automotive industries were also one of the pioneers that utilised different 2D barcodes in a variety of ways, depending on the applications and/or the operating environments. At present, the 2D-symbology-like barcodes used in the Kanban system

Fig. 2.5   An example of Kanban with 2D-symbology-like barcodes (adapted from Figure 1.2-1, p. 10, in [13]).

JAMA Label/Kanban
(Japan & USA)

Fig. 2.6   A recent example of Kanban with 2D barcodes (adapted from [14]).

have been replaced with 2D symbology. Figures 2.5 and 2.6 show early Kanban and recent Kanban, respectively.

Increased data capacity allows some 2D barcodes to have numerous other advantages such as an error detection and correction capability, multilingual encoding and the use of different symbol versions according to the data required. These features enable such 2D barcodes to work as a robust portable database, unlike traditional 1D barcodes, which work only as an index to a backend database. This provided various industries with time-and-cost-effective solutions for managing a number of parts and items without accessing a database, which resulted in their adoption regardless of the space restriction [15]. The development of various application standards in the automotive industry has also encouraged market acceptance for 2D barcode technology. To respond to this demand, around 40 different 2D symbologies have already been developed and the total number is increasing steadily.

Recently, there has been a rapid increase in the number of mobile phones that are equipped with digital cameras [16]. The camera function of a mobile phone enables it to interact with physical objects, including 2D barcode [17]. This opens the possibility of new applications based on a combination of 2D barcode technology and camera-equipped mobile phones.

In addition to 1D and 2D barcode, three-dimensional (3D) barcode also exists today. Three-dimensional barcode is directly marked on the surface of materials such as plastic, rubber and metal. A variety of marking techniques are used for encoding data; these include laser marking, dot-impact marking, thermal marking and sandblast marking. These 3D barcodes can be used where printed labels will not adhere or will be destroyed by a hostile or abrasive environment such as under water. Therefore, 3D barcodes have come into the spotlight as a means to improve traceability; this feature allows manufacturers to manage a product throughout its life, from its production through to disposal [13]. The demand for 3D barcode has escalated, especially in the USA. Associations and organisations including the National Aeronautics and Space Administration (NASA), the Air Transport Association (ATA) and the Automotive Industry Association Group (AIAG) have already developed 3D barcode systems, which are currently in use [13].

2.3   Barcode systems

2.3.1   One-dimensional barcode systems and symbologies

A traditional linear barcode, known as a 1D barcode, is a binary code (i.e. 1s and 0s). The lines and spaces are of varying thicknesses and printed in different combinations [4]. The data, namely an array of 1s and 0s, are encoded in such lines and spaces and the resultant symbol carries the data. The original data can be retrieved by using a scanning device that moves perpendicularly across the lines and spaces. A 1D barcode symbol is usually composed of start and stop characters, data characters, with or without gaps between the combinations of lines and spaces and parity characters. A quiet zone precedes and follows a 1D barcode.

One-dimensional barcode system

One-dimensional barcode is one of the most widespread automatic-identification, keyless-data-entry, technologies. Although there are some exceptions, most 1D barcodes encode a limited number of numeric or alphanumeric characters, with the possible inclusion of some special characters. Consequently, as already noted, 1D barcode works as an index to a backend database. That is, pricing information is not directly encoded in 1D barcode. Instead, information about the product to which the barcode is attached is stored in and retrieved from the database or store controller within the 1D barcode system.

When an index is input at the POS register, the price of the product is retrieved from the database. It is known as a price lookup (PLU) system. Sales prices are often changed

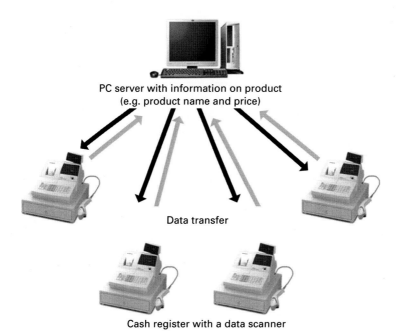

PC server with information on product
(e.g. product name and price)

Data transfer

Cash register with a data scanner

Fig. 2.7 Data flow in the PLU system.

for commercial reasons; to alter the pricing information in the database, the PLU system updates the price of a product without the manual replacement of a price tag attached to the product. Price lookup, together with POS, enables the real-time management of cash flow and inventories, thus enhancing the efficiency and accuracy of sales management. Figure 2.7 shows the data flow in the PLU system.

The benefits of using 1D barcode systems are summarised below.

(i) Fast and accurate keyless data entry is achieved. For example, the probability that Code 39 will be misread is 1 in 3×10^6 characters [10].
(ii) Little employee training is required.
(iii) The numbers encoded into 1D barcode are globally unique. It is advantageous in terms of sales and inventory management.
(iv) Real-time information can be obtained. This is also useful for both sales and inventory adjustment [18].

One-dimensional barcode symbologies

Five different types of 1D barcode have been approved as International Organization for Standardization (ISO) standards among the numerous 1D barcodes that have been developed so far: Code 128, EAN/UPC, Code 39, Interleaved 2 of 5 and the linear symbols of the GS1 DataBar family.[3]

---

[3] The GS1 DataBar, previously called Reduced-Space Symbology (RSS), has also been approved as an ISO international standard. The GS1 DataBar family includes both 1D and 2D barcodes. The symbols of the GS1 DataBar family share common features and are described in Section 2.4.1.

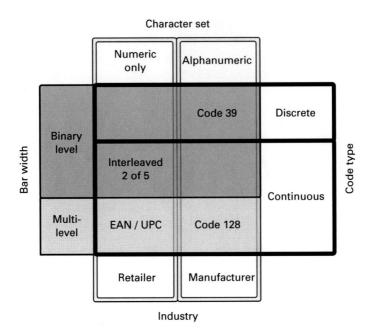

Fig. 2.8   Classification of 1D barcodes. The eight labels around the central square show the categories into which 1D barcodes can be divided.

One-dimensional barcodes can be divided into six different categories by classifying them in terms of bar-width level, code type and the character set used for encoding data. The character set may indicate the groups interested in a particular barcode. For example, 1D barcodes that encode only numeric characters (i.e. EAN/UPC) are often used by retailers, grocery stores and wholesalers. Manufacturers may prefer to use Code 39 and Code 128, owing to their need for encoding alphanumeric characters. Figure 2.8 presents the categories to which each barcode belongs, in terms of bar-width level, code type and character set.

There are two types of bar-width level, *binary level* and *multi-level*. With binary level, combinations of two types of bar width, namely, narrow-bar and wide-bar, compose a barcode. The wide-bar-width to narrow-bar-width ratio typically ranges from 2 to 3. Interleaved 2 of 5 and Code 39, shown in Figure 2.8, are examples of binary-level barcode. Multi-level or (n, k) symbol codes have more than two possible choices for the widths of the bars and spaces. The standard is symbol codes that have four different types of bar width. When being referred to as an (n, k) symbol, n and k indicate the number of basic width units, or *modules* (see Figure 2.11), that compose a data character and the number of bar-width variations, respectively. The EAN/UPC code and Code 128 are categorised as multi-level barcodes. The values of n and k for EAN/UPC equal 7 and 4, respectively, whereas the value of n for Code 128 is 11 with the same bar-width variation (i.e. 4).

In general, the read rate of binary-level barcode is higher than that of its multi-level counterparts, simply because of its easier bar-width detection. Less variation means easier recognition. Conversely, when a multi-level barcode and a binary-level barcode encode identical data, the former often requires less space. That is, multi-level barcode can encode more data than its binary counterparts within a given space. Binary-level barcodes may be the right choice for applications where higher data accuracy is required, whereas multi-level barcodes may be preferred when space is at a premium.

Barcodes can also be divided into *discrete* and *continuous* code types (see Figure 2.8). A space or a gap exists between each code character in a discrete code. Being sandwiched by adjacent gaps, each code character is distinctively identified. A continuous code does not employ such gaps and, as a result, creates a barcode with less width. The continuous code type has a higher data density than the discrete code type within a given space, which is advantageous. However, it is more difficult to print owing to its high data density. The reverse is true for the discrete code type.

Interleaved 2 of 5

Interleaved 2 of 5 is a binary-level continuous barcode. It encodes only numeric character sets (i.e. 0 to 9).

An interleaved 2 of 5 symbol includes a quiet zone, the start pattern, the data character, the stop pattern and a trailing quiet zone. The start pattern consists of a narrow bar, a narrow space, another narrow bar and narrow space, in that order, while a wide bar, a narrow space and a narrow bar comprise the stop pattern.

A data character is made up of either five bars or five spaces, two of which are wide and three narrow in both cases. For example, a combination of two wide bars and three narrow bars comprises a data character. The same applies to the next data character, which is a combination of spaces. The bars and spaces are interleaved together so as to generate a bar and a space alternately. The resultant barcode is an interleaved 2 wide bars out of 5 bars (2 of 5) code. Figure 2.9 shows the data structure of Interleaved 2 of 5.

The number of data characters must be even since every two data characters must be interleaved. However, there is no limitation to encoding the necessary data, provided that sufficient space is available: if there are an odd number of digits, a leading zero may be added.

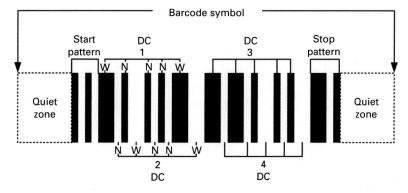

Fig. 2.9   Structure of the Interleaved 2 of 5 barcode: DC, data character; W, wide bar; N, narrow bar.

Since Interleaved 2 of 5 is a variable-length symbology, incorrect data due to an only partial scan may be retrieved. A partial scan has a high probability of decoding a part of a symbol as a valid but short Interleaved 2 of 5 symbol. To prevent such partial scans of long symbols, *bearer bars* should be added to the top and bottom edges of the symbol in the scanning direction. The bearer bars must touch the top and bottom of all the data bars. When a partial scan of a symbol occurs, the scanning beam will cross the bearer bar(s), which prevents the symbols from being decoded. An alternative way of avoiding a partial scan problem is to use Interleaved 2 of 5 symbols in a fixed-length application. The optional checksum is calculated, using a weighted modulo 10 scheme. The last character in the symbol is used for the checksum. When the check character is added, a zero should be placed at the leftmost end of the data area, since Interleaved 2 of 5 must have an even number of data characters, as already mentioned.

Code 39

Code 39 (three–nine) is a binary-level discrete barcode. It encodes alphanumeric character sets. The Code 39 character set includes the 10 digits 0 to 9, the 26 uppercase letters A to Z and seven special characters, −, ., *, $, /, +, % and a space character. It is a variable-length symbology and can be extended to encode the complete 128 American Standard Code for Information Interchange (ASCII) characters by using a two-character coding scheme. Code 39 is widely used in industries and is mandated for some automotive industry and US Department of Defense labels, owing to its ability to handle alphanumeric data.

Code 39 consists of a quiet zone, the start character *, the data characters, the stop character * and a trailing quiet zone. Thus the asterisk is used as both start and stop character. A data character is made up of five bars and four spaces, making a total of nine elements. Each bar or space is either wide or narrow. Three of these nine elements are wide and this is where the code got its name. The code is also known as Code 3 of 9. Its data structure is presented in Figure 2.10.

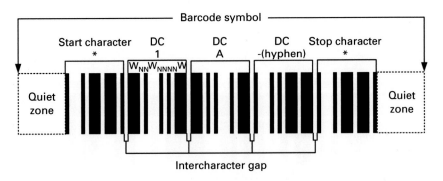

Fig. 2.10   Structure of Code 39: DC, data character; W, wide bar; N, narrow bar.

Regardless of the character set used, in other words, whether numeric or alphanumeric data are encoded, Code 39 requires the same amount of space. It requires substantially more space than other 1D barcodes, since each data character is separated from its neighbouring characters by an intercharacter gap to prevent short reads. As a result, a wider space is required to accommodate Code 39.

Code 39 does not normally include a checksum since the code inherently checks itself based on its code structure: any outcome that does not match the 3-of-9 format is discarded as a wrong decoding. However, it is possible to calculate a checksum with a modulo 43 scheme when required. Each character is assigned a numeric value in addition to its data value, to be used for calculating the checksum. The value of each data character is summed up and divided by 43.

Code 128

Code 128 is a multi-level continuous barcode. The Code 128 character sets includes the 10 digits 0 to 9, both uppercase and lowercase letters (i.e. 'A' to 'Z' and 'a' to 'z') and all standard ASCII symbols and control codes. It is a variable-length very-high-density alphanumeric barcode, designed to encode the full 128 ASCII characters. Unlike Code 39, Code 128 does not require a wider space to encode a given set of data, owing to its use of a unique compression technique. Indeed, it requires the least amount of space to encode data of six characters or more of all the 1D symbologies [8]. Figure 2.11 presents the data structure of Code 128.

The symbol includes a quiet zone, a start character, the data characters, a checksum, the stop character and a trailing quiet zone. Each character consists of three bars and three spaces, varying in width between one and four modules. The total number of modules in each character is 11. However, there is an exception. The stop character is made up of four bars and three spaces, using 13 modules. The bars always consist of an even number of modules and the spaces always consist of an odd number of modules. This gives the basis for a character-by-character consistency check during the scanning process. Code 128

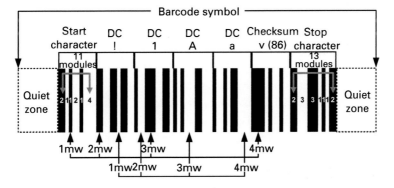

Fig. 2.11   Structure of Code 128: DC, data character; mw, module width.

includes three character sets, A, B and C. Three different start codes are used to indicate which subset will be used. Each can encode the following characters, respectively:

 (i) Code A – the standard ASCII symbols, digits, uppercase letters and control codes
 (ii) Code B – the standard ASCII symbols, digits, upper and lower case letters
(iii) Code C – a set of 100 digit pairs from 00 to 99. By compressing two digits into a character, this provides doubled density.

Using the special characters CODE and SHIFT, code sets can be shifted within a symbol. While CODE shifts the code sets for all subsequent characters, SHIFT changes only the character set of the character immediately following. In addition, SHIFT works only between Code A and Code B.

Code 128 has a checksum, which is calculated using a modulo 103 scheme.

EAN/UPC

The European Article Numbering (EAN) code format was developed using the format of the Universal Product Code (UPC), aiming at an international use of the code. The EAN code is regarded as a superset of UPC. Currently two versions of EAN and five versions of UPC exist: EAN-13 and EAN-8 and UPC-A, UPC-B, UPC-C, UPC-D and UPC-E, respectively. The EAN and UPC symbols are multi-level, continuous barcodes. Only numbers (i.e. 0 to 9) can be encoded. All the characters are fixed in length. However, the length of each symbol depends on the code type and version chosen. It should be noted that versions B, C and D of UPC are not widely used, and so our further description focuses on EAN-13, EAN-8, UPC-A and UPC-E.

EAN-13

The data characters of an EAN-13 symbol, which are enclosed by two guard patterns, namely, the left-hand guard pattern and the right-hand guard pattern, can be divided into two halves, separated by a centre guard pattern.

There are six digits on each side of the centre guard. The last digit of the data characters in the right half of the symbol is the checksum. The data characters surrounded by the guard patterns, quiet zones outside the barcode and human-readable digits below the barcode compose an EAN-13. An EAN-13 data character (i.e. a digit) is represented by seven modules consisting of two bars and two spaces. The widths of the bars and spaces varies between one and four modules. There are 13 data characters in total. The detailed structure of EAN-13 is presented in Figure 2.12.

The data characters in the right-hand half of a symbol are encoded using a *right-hand encoding pattern*, which is an *even-parity* pattern. Conversely, the data characters in the left-hand half can be encoded using two different *left-hand encoding patterns*, one of *odd parity* and one of *even parity*. Three out of the six digits in the left-hand half are encoded using odd parity, and even parity is used for the other three digits.[4] Consequently, the encoding pattern used for any particular digit depends on its location and on whether

---

[4] An exception is a barcode whose first digit is zero. An odd-parity encoding pattern is used for all six digits in the left-hand half of the symbol. In fact, this is actually a UPC code, not an EAN-13 code.

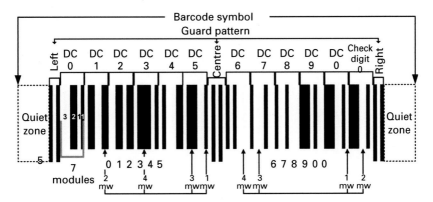

Fig. 2.12   Structure of EAN-13: DC, data character; mw, module width.

odd parity or even parity is used when the digit is within the left-hand half. The digit located at the very beginning of the barcode symbol indicates the combination pattern of odd and even parity for the left-hand half. For example, the '5' at the left in Figure 2.12 indicates that the digits in the left-hand half are encoded using odd, even, even, odd, odd and even parity in sequence; see Figure 4.12(b).

All the data characters on the left-hand side start with a space, whereas all the data characters on the right-hand side start with a bar. The encoding pattern of the right-hand side is designed to be exactly the same as the odd-parity pattern of left-hand encoding but with the '1's and '0's reversed. Furthermore, an even-parity pattern of left-hand encoding and the corresponding right-hand encoding are symmetrical. This feature of the encoding format of EAN-13 enables the symbol's omnidirectional reading.

An EAN-13 barcode is divided into four areas: the number system, the manufacturer code, the product code and the check digit. The first two (or three) digits of the number system are flag characters that identify the country (or economic region) numbering authority which assigned the manufacturer code. In the example in Figure 2.12, the initial number 50 indicates that the United Kingdom (UK) is the numbering authority country that assigned the manufacture identification (ID). The country ID is followed by a five-digit manufacturer ID and a product item ID. The last digit is the checksum and is used to detect and prevent symbol decoding errors. The checksum is calculated using a weighted modulo 10 scheme (see Equation 4.5).

UPC-A

The UPC-A code, which is a subset of EAN-13, is the standard version of the UPC symbols and is widely used in the USA and Canada. Despite its name 'Universal Product Code', the use of UPC-A is limited. It encodes 12 digits. The first is the UPC number system digit, which indicates the type of product (e.g. 0 for groceries and 3 for drugs) [8]. As UPC-A is designed to be used mainly in North America it does not include country codes; otherwise, UPC-A is identical to EAN-13. Any readers who are capable of decoding EAN-13 should be able to read UPC symbols, although the reverse is not necessarily true. Figure 2.13 shows the data structure of UPC-A.

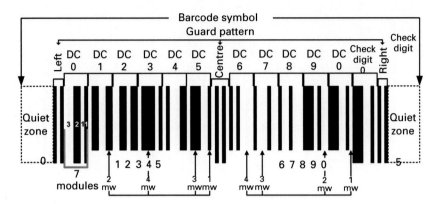

Fig. 2.13   Structure of UPC-A: DC, data character; mw, module width.

Any EAN-13 symbol that starts with the digit 0 is, in fact, a UPC-A symbol, as mentioned earlier. All six digits enclosed between the left-hand and centre guard patterns of EAN-13 symbols are encoded with odd parity only when the first digit is zero. As this implies, UPC-A barcode symbols use exclusively odd parity to encode data in the left-hand half of the symbol.

Using a fixed laser scanner, UPC-A has a first read rate[5] (FRR) of 99% and a substitution-error rate of less than 1 error in 10 000 scanned symbols [8].

EAN-8 and UPC-E

The EAN-8 and UPC-E codes are shortened versions of EAN-13 and UPC-A, respectively. Both EAN-8 and UPC-E were introduced for use on small products, where bigger versions are impracticable.

There is no difference in the data structure between EAN-8 and EAN-13 except for the number of data characters. An EAN-8 code consists of two (or three) digits that indicate the country (or economic region) numbering authority, five digits (or four when the country code has three digits) that indicate the manufacturer and product code and a checksum.

The data structure of UPC-E is different from that of UPC-A. No centre guard pattern exists in a UPC-E symbol. The data characters are enclosed between a left-hand guard pattern and a right-hand guard pattern. The left-hand guard pattern consists of two bars whereas the right-hand guard pattern is made up of three bars. Furthermore, UPC-E does not include a product-type code or a checksum. The UPC-E code is a zero-suppression version of UPC and therefore is also known as zero-suppressed. The data structures of EAN-8 and UPC-E are presented in Figure 2.14.

The checksums for EAN-8 and UPC-E are computed in the same way as for EAN-13 and UPC-A.

---

[5] A measure of a symbology's ability to read the code at the first attempt. It is the ratio of the number of times a good read occurs on the first try divided by the total number of attempts.

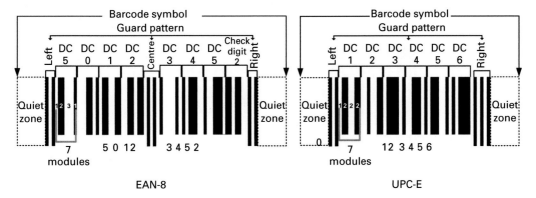

Fig. 2.14 Structures of EAN-8 and UPC-E: DC, data character.

2.3.2 Two-dimensional barcode systems and symbologies

Two-dimensional barcode system

Two-dimensional barcodes combine the advantages of every 1D data-entry method, including portability, readability and excellent information-carrying capability in a small area [4]. Owing to their substantial data-carrying ability, they are sometimes called portable data files or even portable databases. Since a 2D barcode carries data within itself, there is no need to access a backend database in 2D barcode systems. As such, 2D barcodes provide an ideal solution for certain industries (e.g. electronic device manufacturers and automotive manufacturers). Accessing databases for each item is time-consuming and guaranteeing continuous connectivity for a system that does this increases costs. Systems that use 2D barcodes reduce such time and cost.

Industries such as healthcare and electronics also showed an early interest in 2D barcode technology because of the shortage of space on their products. This technology can provide a superior solution for applications where the use of a backend database is impossible or impractical, high-speed processing is required and/or space is a problem. Examples of such applications are on labels on goods in an assembly line, in fast package sorting at an airport and in pharmaceutical labelling in manufacturing, lot-tracking, quality control and so on, in situations where there is limited space.

Although there are some exceptions, 2D barcodes can be classified into two broad categories by the way in which data are encoded and presented, namely, stacked codes and matrix codes. Also known as multi-row codes, stacked codes refers to symbologies made up of a series of 1D barcodes stacked on top of each other. The resulting barcode is two dimensional. Each row of the stacked barcode encodes the data in a series of bars and spaces of varying widths [15], in the same way that 1D barcodes do. Stacked barcodes, however, can contain much more data than their 1D counterparts. However, the use of stacked 2D barcodes also increases the space used to encode the data.

In a matrix code the data are encoded into a matrix format, which is made up of black square cells and white square cells within a matrix. In contrast with stacked codes, matrix

Fig. 2.15   An example of a stacked code (left) and a matrix code (right).

codes can encode a much greater amount of data and in a much smaller space. Hence matrix symbologies can be regarded as true 2D codes [15]. Examples of stacked code and matrix code are presented in Figure 2.15.

The demand to encode more data in a limited space and/or to develop systems that can operate without accessing a backend database has led to the invention of about 40 2D barcodes. Stacked 2D barcodes were first developed mainly to respond to the demand for greater data capacity; they were followed by matrix codes, which were also developed to handle space limitations.

In addition to their symbol structures, a major difference between the stacked and matrix symbologies is the scanners used. Stacked 2D barcodes require specialised laser-based scanning devices, such as raster scanners, that are capable of reading in two dimensions (i.e. vertically and horizontally) simultaneously [15]. A raster scanner operates in the same way as a 1D laser scanner, emitting the same type of light; however, it is capable of oscillating the beam in two dimensions.

Although it is less time efficient, stacked 2D barcode can be read with a traditional 1D scanner. This explains the early prevalence of stacked 2D barcodes as compared with matrix 2D barcodes. Figure 2.16 presents scanning patterns for the raster scanners and laser scanners generally used for scanning 1D barcodes. Raster scanners provide a viable solution for reading stacked 2D barcodes and are usually equipped with the decoding technology needed to decipher the code when it is damaged [15]. However, they are not capable of deciphering matrix 2D barcodes as their decoding process involves more complicated translation.

To read matrix 2D barcode, charge coupled device (CCD) imagers are usually used. A CCD, which is a semiconductor with a plurality of light-sensitive areas, can have either a single linear set of light-sensitive elements or a light-sensitive area arranged in a two-dimensional array. The former is called a *linear* or *line* (sometimes, *one-dimensional*) *sensor*, whereas the latter is known as an *area sensor*. A scanner equipped with an area sensor is necessary to read 2D barcodes; a linear sensor scanner can only read 1D barcode. Figure 2.17 shows a CCD scanner, an area sensor and a linear sensor.

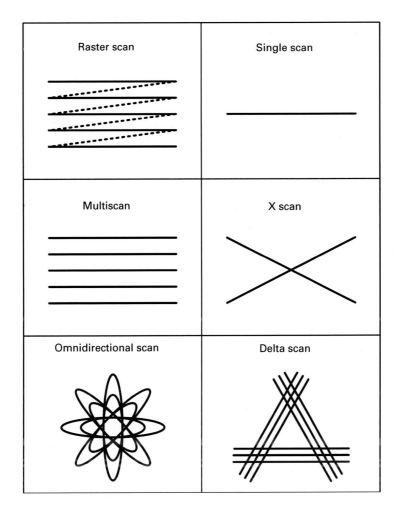

Fig. 2.16   Laser scanner patterns.

Each type of 2D barcode has evolved independently to meet the requirement for certain groups, in contrast with the 1D barcode system where standardisation of the symbol and the code structure was the initial goal. For example, MaxiCode is of fixed size (1 inch × 1 inch) and can encode only up to 93 characters (alphanumeric) [13]. (The symbol description is provided later in this chapter.) It appears to be very limiting. However, the code was designed as a best fit to its application [4]. It was developed by the United Parcel Service (UPS) for shipping applications where high-speed sorting is required. If the central bull's eye were of variable size, searching for it would be much more complex. In addition, if the cell size of symbols were smaller then the individual hexagons would not be resolvable without an increase in the reader resolution. But, for example, doubling the resolution for the same belt width would increase fourfold the number of samples per second [4].

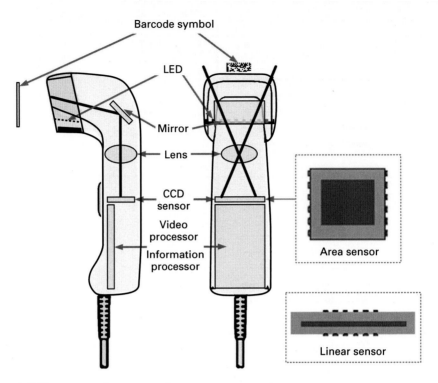

Fig. 2.17   A CCD scanner and two types of sensor.

A company may well choose a stacked 2D barcode for a system unless space is a problem, as with semiconductor manufacture for which a matrix code is preferable. Two-dimensional systems work effectively as long as groups with a common interest use the same code structure. This may explain the trend of 2D barcode development.

The information encoded in 1D barcode is vertically redundant. Even if part of a symbol is damaged by defects such as a smudge, spot and tear, the original information can be accurately retrieved from the other part of the bars and/or spaces. Furthermore, human-readable information (e.g. alphanumerical characters) is printed beneath 1D barcode symbols to allow the manual input of data if a decoding error should occur. Unlike 1D barcode, each small cell of the matrix format of 2D barcode symbols has information, and there is no redundancy to overcome printing defects. Since, in general, the amount of data encoded in a 2D barcode symbol is not small, printing human-readable data beneath the symbol is impractical. If a few cells are damaged, the data encoded within the 2D barcode symbol will be lost. Additional techniques should be applied to prevent this.

Nearly all 2D barcodes use an error detection and correction technique. For example, all the ISO standard 2D barcodes use Reed–Solomon coding for error correction.[6] With

---

[6] Data Matrix ECC levels 000 to 140 use convolutional code error correction. These are still valid. However, ECC 200, which uses the Reed–Solomon code, is in common use.

data detection and correction features, 2D barcodes can substantially improve reading robustness and, as a result, they achieve high readability. This is especially true when 2D barcodes are decoded with a dedicated scanner. Some 2D barcode scanners can detect, locate and decode the 2D barcodes placed on moving targets such as luggage on a conveyor belt. Increased data capacity also allows 2D barcodes to have additional useful features. These include multilingual data coding, a structured append function (described in detail in Chapter 5) and encryption protection for security.

Two-dimensional barcode systems, which are the integration of robust 2D barcode symbols and high-quality scanners, can offer a variety of robust and reliable applications according to user needs. Initially, the use of 2D barcode was limited to within an organisation, manufacturer or groups with a common interest. Lately, however, 2D barcodes have prevailed into our everyday lives. The 2D barcodes that have replaced stamps for postage are one such example.

In November 2000, in Japan, J-Phone[7] and Sharp Corporation released J-SH04, the world's first commercial camera phone with an integrated CCD sensor. This invention has sparked interesting applications that can integrate barcode and camera phone technology. With software for decoding, a camera phone can perform scanning and decoding tasks on the device itself. If the right decoding software is installed, the user of the camera phone can decode the corresponding 2D barcodes.

The number of camera phone users has increased rapidly and the research firm Gartner predicts that this global trend will continue, replacing mobile phones without a camera [19]. Mobile phones and camera phones have become part of our everyday lives, being always within reach and providing seamless network connectivity not only for human-to-human communication but also for human-to-technology interaction. Camera phone users can talk, take photos, use the *short message service* (SMS) and the *multimedia messaging service* (MMS), send emails, and link their phones to the Internet, regardless of time and location. Furthermore, they can perform additional tasks such as Web browsing, on-line shopping and updating their Web logs (known as blog) using the Internet connection established via a camera phone.

Two-dimensional barcodes can be used as a tag to connect camera phones and their users to the Internet or to other technologies and devices such as a plasma display, enabling interaction between them. In other words, 2D barcodes can link the physical and digital worlds. This idea attracts researchers whose interest is in developing an environment where the physical world and digital worlds are integrated so as to provide humans with additional information or better services by preserving continuous connectivity. The former concept is known as *augmented reality* (AR). The latter concept, known as ubiquitous computing, has come into its own in recent years. Some researchers have developed their own 2D barcodes to make their novel ideas reality [17], [20].

Two-dimensional barcode symbologies
Nearly 40 2D barcodes have been invented, some for industrial use and some as a tag to link the physical and digital worlds. Six are established as an ISO standard. These

---

[7] One of the mobile phone operators in Japan at the time and now Softbank Mobile Co.

include PDF417, MicroPDF417, Data Matrix, MaxiCode, QR Code and the 2D symbols of the GS1 DataBar family (known as RSS-14 Stacked Symbol). The specifications of PDF417, MicroPDF417 and MaxiCode are provided below. Data Matrix and QR Code are also well known as 2D barcodes for mobile applications. Their specifications are included in Chapter 3.

PDF417

Portable Data File (PDF) 417 was invented by Ynjiun Wang in 1991 at Symbol Technologies Inc. It was one of the earliest 2D barcodes on the market, probably owing to the fact that PDF417 symbols can be read by not only CCD cameras but also linear scanners or laser scanners, which were already in use. The flexibility in shape of PDF417 and its high data detection and correction capability also attracted the market. This is the most widely used stacked 2D barcode and is popular globally [13]. In 2000, PDF417 was approved as an ISO standard (ISO/IEC 15438).

This barcode has a multi-row, continuous, variable-length symbology. As the name suggests, PDF417 has high data capacity and is capable of encoding numeric, alphanumeric and binary data. A PDF417 symbol includes the start pattern, the left-hand row indicator, the data characters, the right-hand row indicator and the stop pattern. The symbol is surrounded by a quiet zone. The left-hand and right-hand row indicators contain the format information such as the row identifier, the total number of rows and columns and the error correction level.

The basic unit of PDF417 is a codeword that starts with a bar and ends with a space. A codeword of the PDF417 symbol is made up of 17 modules arranged in four bars and spaces of varying width; this gives rise to the name of the symbology. The sequence of the widths of the bars and spaces used in a codeword, in other words, the sequence of the numbers of modules used for the bars and spaces is the value of the codeword. The structure of PDF417 is presented in Figure 2.18.

Every PDF417 symbol consists of a stack of three to 90 rows. Each PDF417 row can contain from one to 30 data symbol characters. All rows in a symbol have the same number of codewords. The first codeword in the data area indicates the total codeword number and the last few codewords in the last row are used for error detection and correction.

There are three mutually exclusive sets of symbol patterns, or *clusters*. Each cluster, namely, Cluster 0, Cluster 3 and Cluster 6, has 929 distinct codeword patterns. In each cluster the first 900 codewords (from 0 to 899) are used to represent the data, and the rest (from 900 to 928) are used for special functions. Codewords from the same cluster are used for encoding in each row. The three clusters are repeatedly used in sequence (e.g. Cluster 0, Cluster 3, Cluster 6, Cluster 0, Cluster 3, . . .). Since adjacent rows use different clusters, the decoder can detect a cluster and then stitch it into the appropriate location if the scanning line is crossing row boundaries. Figure 2.19 shows the patterns of bars and spaces used to encode 0 in each cluster.

The PDF417 symbology has three predefined modes and nine reserved modes for encoding data. The predefined modes are binary, EXC (i.e. text) and numeric. In the binary mode, approximately six characters can be encoded into five codewords. In the EXC mode, the alphanumeric data can be encoded in double density (i.e. two characters

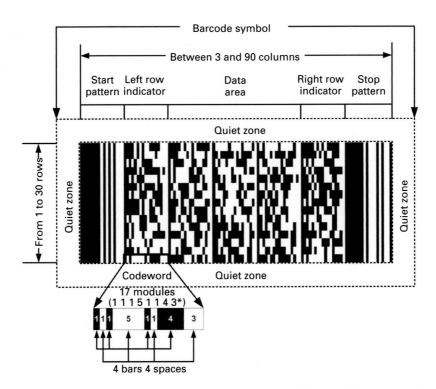

Fig. 2.18   Structure of PDF417. The sequence of the numbers of modules in each bar or space corresponds to a codeword.

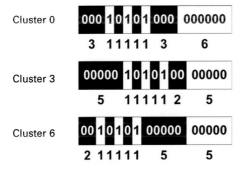

Fig. 2.19   PDF417 bar-and-space patterns encoding the value 0 in the three different clusters, each of which has 17 modules.

per codeword). In numeric mode, the numeric data can be packed into almost triple density (i.e. three characters per codeword). This allows one PDF417 symbol to encode up to 1108 bytes, 1850 ASCII characters or 2725 digits, depending on the selected *data correction mode*. Three different data compaction modes, i.e. the *text compaction mode*, *byte compaction mode* and *numeric compaction mode*, define the mapping between the codeword values and the decoded data.

Table 2.1. PDF417 error correction level

Error correction level	0	1	2	3	4	5	6	7	8
Number of error correction codewords	2	4	8	10	16	32	64	128	256

Every PDF417 symbol uses the Reed–Solomon error correction technique, which allows the symbol to tolerate some damage without data loss. There are nine error correction levels, ranging from 0 to 8. The error correction level depends on the amount of raw data, the symbol size and the environment in which the symbol will operate. A higher level of error correction may be required in harsh environments. Every PDF417 symbol includes at least two error correction codewords, and up to 510 error correction codewords can be added to a symbol. Table 2.1 presents the error correction levels of the PDF417.

As the number of rows and their lengths are selectable, the aspect ratio of a PDF417 symbol can be varied to suit the spatial requirements for printing [4]. The PDF417 code also has a structured append feature called MacroPDF. Thus, if the amount of data to be encoded exceeds the maximum data capacity of PDF417 or if a PDF417 symbol is too large to be placed within the available space, it can be divided into several smaller PDF417 symbols that can logically link. Each MacroPDF segment has a file identifier (ID) with a unique index in its control block to ensure that it is reassembled in the correct order regardless of the reading order of segments. If a segment has a different file ID, this indicates that it does not belong to that particular barcode sequence. In a very clean environment, where a PDF417 symbol is less likely to be damaged, the use of TruncatedPDF417 can provide a space-efficient solution.

MicroPDF417

MicroPDF417, which is a derivative of PDF417, is a multi-row symbology developed by Symbol Technologies Inc. It is also an established ISO standard (ISO/IEC 24728) and was designed for applications where improved space efficiency is required [1]. A limited set of symbol sizes is available. MicroPDF417 has four different *column versions*, 1, 2, 3 and 4 with limited variations in the number of rows. Column version 1 can have between 11 and 28 rows, and between 8 and 26, 6 and 44, and 4 and 44 rows are available for column versions 2, 3 and 4, in that order. A MicroPDF417 consists of a data area and *row address patterns* that enclose the data area. Column versions 3 and 4 have an additional row address pattern in the centre of the symbol.

The basic unit of MicroPDF417 is again the codeword. The codeword structure of MicroPDF417 is the same as that of PDF417. A codeword is made up of four bars and four spaces with varying widths to make a total of 17 modules as before. Again it starts with a bar and ends with a space. Figure 2.20 presents the code structure of MicroPDF417. Data are encoded in a MicroPDF417 symbol in the same way as in PDF417 symbols.

MicroPDF417 uses three different *compaction modes*, binary, text and numeric. The symbol allows up to 150 bytes, 250 alphanumeric characters or 366 numeric digits to be stored by specifying one of the three compaction modes. The error correction level of

Fig. 2.20   Structure of MicroPDF417.

MicroPDF417 is fixed by the numbers of columns and rows. The number of correctable codewords is in the range seven to 50. Detailed information on all four column versions, i.e. the number of rows, corresponding error correction capability and data capacity for the three compaction modes is presented in Tables D.1 and D.2 in Appendix D.

MaxiCode (UPSCode)

MaxiCode, originally called UPS Code, was created by UPS in 1992 for the high-speed sorting and tracking of unit loads and transport packages. It is a fixed-size 2D barcode with limited data capacity, which results in a reduction in decoding time. Although the data capacity of MaxiCode is less than that of other 2D barcodes, it is sufficient to handle applications and contributes to minimising the decoding time. MaxiCode is 15% denser than square dot codes owing to its symbol structure, and approximately 100 ASCII characters can be encoded in a 1 inch × 1 inch symbol. MaxiCode is ideal for applications where the label is on a moving package, or its orientation is random, or space is limited, or the scanner is placed so that a large view of the package is taken [4]. MaxiCode was

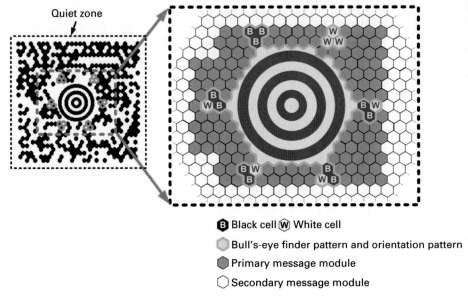

Quiet zone

B Black cell **W** White cell

Bull's-eye finder pattern and orientation pattern

Primary message module

Secondary message module

Fig. 2.21   Structure of MaxiCode.

approved as an ISO standard (ISO/IEC 16023) in 2000. It was also adopted as a two-dimensional symbol for the shipping, transport and receiving of labels (ISO 51394). The data structure of MaxiCode is presented in Figure 2.21.

One of the most distinctive features of MaxiCode is its unique circular finder pattern made up of three concentric rings called a bull's eye. At six fixed locations outside the bull's eye, six different *orientation patterns* are allocated, enabling the detection of the symbol's orientation. Each orientation pattern is made up of a combination of three modules (i.e. cells), using black, white or both black and white modules. Offset rows of hexagonal modules are arranged around the bull's eye, producing a nominal 1 inch by 1 inch (1.11 inch × 1.054 inch) array of 884 black interlocking hexagons. These hexagonal modules are arranged in 33 rows, each row containing up to 30 modules. A codeword of MaxiCode consists of six modules, usually arranged as three rows × two columns. Up to 144 codewords can be encoded in a MaxiCode symbol.

With six modules representing six bits, MaxiCode can produce 64 different codewords, which enables it to encode a maximum of 256 (i.e. ASCII and Extended ASCII) data characters. Five different code sets, A, B, C, D and E, are used to achieve this. The numeric compaction mode allows MaxiCode to encode nine digits into six codewords. As a result, the maximum data capacity of MaxiCode is 138 digits; up to 93 characters of alphanumeric data can be encoded, and MaxiCode also uses 15 symbol control characters such as shift and latch characters to control decoding devices and manage data.

MaxiCode is also characterised by its two-message format. The data area is divided into two parts, the *primary message* and the *secondary message*. The primary message encodes a postal code, three-digit country code and three-digit service class code in up to

Table 2.2. Numbers of various categories of codeword in SEC and EEC

	SEC	EEC
Total codewords	144	144
Data codewords	93	77
Codewords for error correction	50	66
Correctable codewords	22	30

120 modules (i.e. 20 codewords) whereas the maximum data capacity of the secondary message area is 744 modules (i.e. 124 codewords), used for encoding all the other data.

MaxiCode uses the Reed–Solomon error detection and correction technique, providing two levels of error correction: *standard error correction* (SEC) and *enhanced error correction* (EEC). The SEC level provides 15% error correction capability for misread data and 30% for data that could not be read. The EEC level can achieve 21% and 42% error correction capability for misread data and data that could not be read, respectively. Table 2.2 shows the numbers of data, error correction and correctable codewords for SEC and EEC.

MaxiCode defines seven modes that determine how the data should be encoded in the two-message structure, with different levels of error correction, as shown in Table 2.3.

The first two modes (i.e. Mode 0 and Mode 1) are no longer used. Modes 2 and 3 are for *structured carrier messages* and require specific data in the correct order to produce a scannable symbol. Modes 4 and 5 are used to encode raw data, Mode 5 offers a slightly higher level of error correction. Mode 6 encodes a message (i.e. instruction) on how to program the reader system.

The MaxiCode structured append[8] feature allows up to eight symbols in a structure. To enable the structured append feature, the *sequence indicator* (SI) and the total number of symbols (TS) must be encoded in each symbol. The SI is a one-based index that identifies the position of this particular symbol in the group and the TS indicates the number of total multiple symbols; SI and TS can each be any number between 1 and 8. An efficient reader can buffer the contents of each symbol until all the symbols are read and then retrieve the original data.

The MaxiCode symbol can be read by using a CCD camera or laser scanner. However, the symbol must be printed by a high-resolution printer such as a thermal transfer printer.

2.3.3 Three-dimensional barcode systems and symbologies

Colour 2D barcode symbols use colour elements over their spatial two-dimensional component for encoding data and, accordingly, may be classified as 3D barcode. However, spatial 3D barcodes exist and are used where label printing with ink or paint is not practical or printed labels are not reliable owing to a surrounding harsh environment.

---

[8] The structured append is described in detail in Section 5.3.3.

Table 2.3. MaxiCode encoding modes

Mode	Description
0	Obsolete and replaced by Mode 2 and Mode 3.
1	Obsolete and replaced by Mode 4.
2	Structured carrier message: the primary message encodes the numeric postal code, country code and service code (up to nine digits) and the secondary message encodes additional data.
3	Structured carrier message: the primary message encodes the alphanumeric postal code, country code and service code (up to six characters) and the secondary message encodes additional data.
4	Used for any data with EEC for the primary message and SEC for the secondary message. Up to 84 data characters, automatically split between primary and secondary messages.
5	Used for any data with EEC for both the primary and secondary messages. Up to 68 data characters, automatically split between primary and secondary messages.
6	Reader programming mode: similar to Mode 4, but uses reader control.

In this book, we will regard barcodes that are composed of three spatial dimensions (i.e. x, y, z coordinates) as 3D barcodes. Consequently, two-dimensional barcodes that have additional elements such as colour will be referred to as 2D barcodes with an extra description. For example, if a 2D barcode has a colour element, it will be called colour 2D barcode. A 2D barcode with an animation element will be called animated 2D barcode.

A barcode directly embossed on the surface of a material is regarded as a 3D barcode (or 'bumpy' barcode). Both 1D and 2D barcode formats can be used to create 3D barcode symbols. Instead of visual contrasts between different colours, 3D barcode readers use spatial elements or the bumpy aspect of the symbol to locate symbols, measure the width of bars and spaces (or the size of cells when 2D barcode is used) and decode them. Three-dimensional barcodes can be painted, coated or made a permanent feature of an item, permanently retaining their readability [3] and trackability. Figure 2.22 presents an example of such a 3D barcode.

Direct marking and the resulting 3D barcodes have attracted attention as an environmentally friendly technology because no additional materials are used to generate the symbols and the product is trackable. The reuse and recycling of products has become increasingly important in preventing environmental destruction and protecting our limited resources [13]. A system that can control and manage the lifecycle of each product is essential for promoting product reuse or recycling. Three-dimensional barcodes that can work as a reliable tool to trace the life cycle of products play a vital role in nature conservation.

Marking techniques such as *laser marking* and *dot peen marking* are often used to generate 3D barcode symbols. To read them, special readers and lighting may be required, respectively [10]. Since 3D images cannot be captured by camera phone, 3D barcode technology is outside the scope of this book.

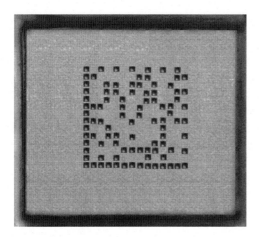

Fig. 2.22   Example of a 3D barcode.

2.4   Emerging technologies and barcode

Since the birth of 1D barcode in the 1970s, barcode technology has evolved continuously. Different types of barcode technologies such as 2D and 3D barcodes have been developed to support a wider range of applications. As discussed in the preceding sections, numerous barcode symbologies, each with unique features, have been introduced. This trend is ongoing and is expected to continue. Instead of competing with each other, different types of barcode systems (1D, 2D and 3D) have co-existed, each opening up its own market. Is there any possibility for this situation to change, for example, owing to the emergence of new types of barcode technologies and applications or to the emergence of technologies other than barcodes?

In 2006 a leading organisation, GS1, that develops and implements global standards for supply chains, announced a global sunrise date of 1 January 2010 when the newly developed GS1 DataBar family symbologies will be available in all trade-item scanning systems. The proliferation of radio frequency identification (RFID) technology, which is often compared with 2D barcode technology, has escalated in recent years owing to its reducing cost and its capability as a ubiquitous computing tool. A short-range wireless technology called *near field communication* (NFC) has also attracted attention as a suitable tool for ubiquitous computing.

The major advantage of barcode technology is its inexpensive operation, using paper and ink. New barcode applications can operate even without paper and ink. These applications use barcode symbols that are digitally displayed on a medium such as a mobile phone display. Direct marking also requires no additional materials and so less waste will be produced, demonstrating its environmental friendliness.

Considering the current trend of these technologies, in this section we discuss first how each barcode system will probably progress, next, how emerging technologies are expected to affect the development of barcode technologies and, finally, how barcode technology may contribute to environmental preservation.

2.4.1   Barcode technology forecast

Although most 1D barcodes are capable of encoding only a limited number of digits, they can be seen as advantageous especially in standardisation. Unlike languages, numbers are commonly used throughout the world; this has simplified standardisation and attracted huge industry support. Thus, the 1D barcode system has resulted in huge success in most interest groups, such as grocery manufacturers, wholesalers and retailers.

Two-dimensional barcode systems can provide a superior solution for applications where the use of a backend database is impossible or impractical, high-speed processing is required and/or space is a problem. Each 2D barcode and its system has evolved independently to provide interest groups with better system management in terms of time and cost, which has resulted in better productivity. This trend is expected to continue in certain industries.

Activities towards the utilisation of 2D barcode have been seen within companies or groups with common interests. However, there has been little interest from retailers, wholesalers and the grocery industry, indicating that the 1D barcode system will progress further. This is mainly because there is no major problem with the current use of standardised 1D barcodes such as EAN13/8, UPC-A/E and UCC/EAN-128[9] [13]. It is nearly impossible to introduce and establish new 2D barcode systems when little demand exists.

Furthermore, the adoption of GS1 DataBar symbologies (formerly called reduced-space symbology or RSS) covers the shortcomings of the existing 1D barcodes in the grocery and pharmaceutical industries, where items are sometimes too small to place these symbols (see subsection 2.3.1). Reduced-space symbologies were developed with the collaboration of EAN International and the Uniform Code Council (UCC) in 1996. Existing 1D barcodes such as EAN and UPC can encode only up to 13 digits (12 digits in the case of UPC), whereas symbologies in the RSS family are capable of encoding the 14-digit GS1 Global Trade Item Number (GTIN). As the name suggests, RSS symbologies are designed to fit a small space without reducing the data capacity of the symbols. This is achieved by a compressed format, owing to which it is also possible to place RSS symbols on the curved surfaces of products such as vials and test tubes that contain drugs. Although the initial purpose of developing RSS symbols was to use them for labelling small and/or non-flat products such as meat, unit doses and agricultural produce, the use of RSS is expected to expand gradually and become widely accepted.

The GS1 organisation was formed by EAN International and over 100 local member organisations, including the UCC from the USA and the Electronic Commerce Council of Canada (ECCC). It was established as a leading global organisation dedicated to the design and implementation of global standards and solutions to improve the efficiency and visibility of supply and demand chains [21]. In fact, GS1 promoted the global adoption and use of the 14-digit GTIN from 1 January 2005. The aim was to speed up data flow in commercial exchanges and to optimise the control of goods throughout the world,

---

[9] UCC/EAN128 is currently called GS1-128. It is the application standard of the GS1 implementation based on Code 128 specification.

by providing a universal standard barcode [21]. In June 2006, GS1 announced a global sunrise date of 1 January 2010 for RSS, following a compelling business case review by a global task force comprising retailers, fast-moving consumer goods manufacturers, pharmaceutical companies, GS1 member organisations (MOs) and trade associations. This means that RSS symbols and GS1 *application identifiers* (AIs) should be available in all trade-item scanning systems from 1 January 2010. In February 2007, RSS was officially renamed GS1 DataBar™.

GS1 DataBar symbologies

In 1999, the symbologies of the GS1 DataBar family (RSS at that time) were standardised as the International Technology Specification (ITS) by the association of Automatic Identification Manufacturers (AIMs). The symbologies were also approved as an ISO standard (ISO/IEC 24724).

The GS1 DataBar family[10] contains three linear symbologies: GS1 DataBar Limited, GS1 DataBar Omnidirectional and GS1 DataBar Expanded. There are three variations of GS1 DataBar Omnidirectional. These are GS1 DataBar Truncated, GS1 DataBar Stacked and GS1 DataBar Stacked Omnidirectional. GS1 DataBar Expanded also has a variation, namely, GS1 DataBar Expanded Stacked. The symbols of the GS1 DataBar family are presented in Figure 2.23; the hierarchy of the family moves from left to right.

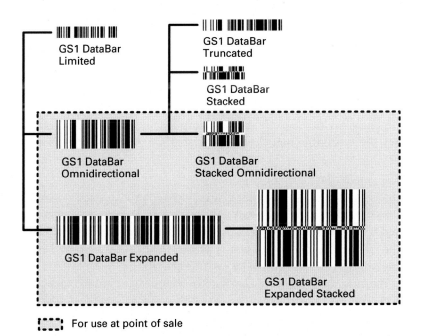

Fig. 2.23   Symbols of the GS1 DataBar family. The data for GS1 DataBar Expanded and GS1 DataBar Expanded Stacked symbols are (01)98898765432106(3202)012345(15)991231; the data for the other symbols are (01)00501234567890.

---

[10] The symbologies of the GS1 DataBar family were used in ISO/IEC 24724, unless otherwise indicated.

Of the seven different formats of the GS1 DataBar family, four are designed specifically for use in retail POS systems. They are GS1 DataBar Omnidirectional, GS1 DataBar Expanded, GS1 DataBar Stacked Omnidirectional and GS1 DataBar Expanded Stacked. The features of these POS symbols are as follows.

GS1 DataBar Omnidirectional

This symbol was developed for use at POS for standard item identification. It encodes the full 14-digit GS1 item identification in a linear symbol that can be scanned omnidirectionally by suitably programmed POS scanners. As mentioned in Chapter 1, the X dimension is the width of the smallest element in a barcode symbol. The symbol size is 33X (this is the height H) $\times$ 93X (this is the length L). When X is 0.17 mm, the smallest symbol size will be 5.6 mm (H) $\times$ 15.8 mm (L).

GS1 DataBar Stacked Omnidirectional

This symbol, which encodes the full 14-digit GTIN, is a variation of GS1 DataBar Omnidirectional, as mentioned earlier. As the name indicates, the symbol format is designed for the fixed-position omnidirectional scanners commonly used in supermarkets. The symbol is typically used for loose produce items such as apples, potatoes and oranges, which can provide a narrow and tall space for labelling. The symbol size is 69X (H) $\times$ 50X (L). When X is 0.17 mm, the smallest symbol size will be 11.7 mm (H) $\times$ 8.5 mm (L).

GS1 DataBar Expanded

This symbol, which can encode up to 74 alphanumeric characters, was developed to provide a variable-measure identification of items such as meat, seafood and delicatessen. The supplementary application identifier (AI) and element strings such as the weight and the 'best before' date are encoded in a linear symbol in addition to the GS1 item identification. It can be decoded omnidirectionally. The smallest symbol height is 34X (H). When X is 0.17 mm, the largest symbol size is 5.8 mm (H) $\times$ 91.9 mm (L).

GS1 DataBar Expanded Stacked

This symbol can also be printed in multiple rows as a stacked symbol, resulting in a GS1 DataBar Expanded Stacked symbol. GS1 DataBar Expanded Stacked has the same characteristic as GS1 DataBar Expanded except where width constraints exist. The minimum height of each row is 34X (H). When X is 0.17 mm, the largest symbol size with six rows is 34.7 mm (H) $\times$ 17.3 mm (L).

The three remaining GS1 DataBar symbols are designed for applications other than POS. They all encode the full 14-digit GTIN in symbols that cannot be scanned omnidirectionally. Charge coupled device (CCD) scanners and laser scanners are used to read them. These symbols are as follows.

GS1 DataBar Truncated

This symbol, which is a variation of GS1 DataBar Omnidirectional, is designed for use on items such as cosmetics and jewellery. The symbol size is 13X (H) $\times$ 93X (L). When X is 0.17 mm, the smallest symbol size is 2.2 mm (H) $\times$ 15.8 mm (L).

GS1 DataBar Stacked

This symbol is a variation of the GS1 DataBar Omnidirectional symbol that is stacked in two rows and is used when the normal symbol would be too wide for the desired application. It is designed for use on small items requiring a narrow and highly truncated format. The symbol size is 13X $(H) \times$ 50X (L). When X is 0.17 mm, the smallest symbol size is 2.2 mm $(H) \times$ 8.5 mm (L).

GS1 DataBar Limited

This is a very small GS1 DataBar symbol format. Its indicator digit must be a '0' or '1'. The symbol size is 10X $(H) \times$ 70X (L). When X is 0.17 mm, the smallest symbol size is 1.7 mm $(H) \times$ 11.9 mm (L).

The symbologies of the GS1 DataBar family are multi-level, continuous and character self-checking linear barcodes. Each GS1 DataBar symbol consists of the left-hand and right-hand guard patterns, the data characters and the finder patterns. The guard patterns consist of two one-module-wide elements forming either a bar and space or a space and bar pair at each end of the symbol. Stacked symbols such as these are found in GS1 DataBar Stacked and GS1 DataBar Expanded Stacked have guard patterns at the end of each row of the symbol.

Every symbol has two or more data characters, each with an (n, k) structure, where n is the number of modules and k is both the number of bars and the number of spaces that compose the data character. Various different (n, k) symbol characters are used for each symbol. However, all can be read bi-directionally. GS1 DataBar Omnidirectional and GS1 DataBar Limited are fixed in length, whereas GS1 DataBar Expanded is a variable-length symbol. The data character values are combined mathematically to form the explicitly encoded data.

The finder pattern is a set of elements (i.e. a combination of bars and spaces) that are set to be identifiable by the decoder, so that the symbol can be recognised and the relative position of the elements can be determined. Each symbol contains one or more finder patterns. The finder patterns also function as the *check character* and/or *segment identifiers*. The code structures of the original members of the GS1 DataBar family, namely, GS1 DataBar Omnidirectional, GS1 DataBar Limited and GS1 DataBar Expanded, are presented in Figure 2.24.

The encodable character set for GS1 DataBar Omnidirectional and GS1 DataBar Limited is 10 digits (i.e. 0 to 9), whereas a subset of ISO 646 can be encoded in GS1 DataBar Expanded. The subset includes upper and lower case letters, digits and 20 selected punctuation characters (!, ", %, &,', (,), *, +, ,, -, ., /, :, ;, <, =, >, ?, _) in addition to the special GS1 function character (i.e. FNC1). The maximum data capacity of GS1 DataBar Omnidirectional and of GS1 DataBar Limited is a 14-digit numeric item identification that represents GTIN. The AI '01' is added to it by decoders. GS1 DataBar Expanded is capable of encoding up to 21 data characters, which can produce 74 numeric and 41 alphanumeric characters. The maximum data capacity of each symbol does not include any encoded FNC1.

Every GS1 DataBar symbol is capable of error detection. The checksums of GS1 DataBar Omnidirectional, GS1 DataBar Limited and GS1 DataBar Expanded are calculated using weighted modulo 79, 89, 211 schemes, respectively.

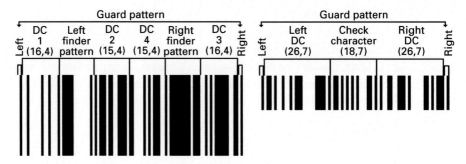

GS1 DataBar Omnidirectional and GS1 DataBar Limited symbol representing (01)00501234567890

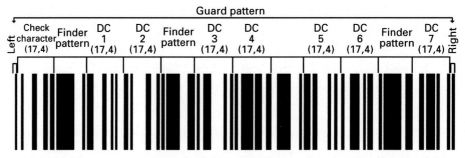

GS1 DataBar Expanded symbol representing (01)98898765432106(3202)012345(15)991231

Fig. 2.24   Structures of the GS1 DataBar family symbologies: DC, data character; (n, k) represents the number of modules n and the number of bars or spaces k in a data character.

Some features increase the usability and usefulness of the GS1 DataBar family symbologies: *data compaction*, *component linkage* and *GS1-128 emulation*. Each GS1 DataBar symbol has data compaction methods that are optimised for the data string to be encoded. For example, GS1 DataBar Expanded uses data compaction methods optimised for specific combinations of application identifiers that are commonly used. The 2D Composite component can be used along with the linear component of GS1 DataBar symbols. Each DataBar symbol contains a linkage flag that when set indicates that the 2D Composite component is adjacent to the linear GS1 DataBar symbol. If the linkage flag is set, the reader will decode the symbol accordingly. GS1-128 emulation enables code readers that are set to the GS1-128 emulation mode to transmit the data encoded in the symbols of the GS1 DataBar family as if the data were encoded in one or more GS1-128 symbols.

The GS1 DataBar symbols are exclusively reserved for encoding identification numbers and data supplementary to the identification. Administration of the numbering system by GS1 ensures that identification codes assigned to particular items are unique worldwide and that they, and the associated supplementary data, are defined in a consistent way. The major benefit for the users of the GS1 system is the availability of uniquely defined identification codes and supplementary data formats for use in their trading transactions. Any member of the GS1 DataBar family can be printed as a stand-alone linear symbol or as a composite symbol with an accompanying 2D Composite component printed directly above the GS1 DataBar linear component.

2D barcode for mobile devices as a ubiquitous computing tool

Considering the proliferation and the usefulness of 1D barcodes, it is unlikely that 2D barcodes will totally replace them. However, this does not depreciate the value of 2D barcodes. For the last few years, 2D barcode technology has been in the spotlight as a ubiquitous computing tool. With the integration of CCD cameras in recent years, mobile phones have become networked personal-image-capture devices. Furthermore, programmable camera-equipped mobile phones (i.e. camera phones) can perform image processing tasks on the phone itself and use the outcome as an additional means of input by the user and a source of context data [17]. A 2D barcode with high data capacity can carry sufficient data within itself to provide meaningful information whenever it is required and/or accessed. The integration of two such portable technologies, namely, 2D barcode technology with camera phone technology can provide simultaneously the data for an application and a device to interpret and implement it, regardless of time and location.

The most well-known application implemented by the use of 2D barcode and camera phone technology is the linking of camera phone to the Internet. A camera phone can be automatically connected to a Web page by capturing a 2D barcode symbol in which its *uniform resource locator* (URL) is encoded, using the built-in camera of the phone. Camera phone users can perform additional tasks using the link that has been established at any time and anywhere in the world. The combination of 2D barcode and camera phone technology can also be used when no network connectivity is available. For example, a camera phone user can create an address book adding and saving contact details of people by capturing each 2D barcode symbol where the contact information is encoded. All that is needed to perform such an application is a camera phone with the appropriate software and 2D barcode symbols that are placed on the media such as business cards and blogs.

These applications are important in that they have opened the door for ordinary users to use 2D barcode technology as one of the ubiquitous computing tools. This is a new type of service, enabled by the recent proliferation of camera-equipped mobile phones and the growing popularity of 2D symbologies, which has potential for further development.

If the proper infrastructures and systems are in place, it is possible to have widespread use of 1D barcodes for mobile applications as they are already printed on billions of product packages worldwide. There are some ongoing projects to implement 1D barcode systems in the context of ubiquitous computing. Owing to their limited data capacity, reading requirement, symbol shape and/or space requirement, there might be limitations to using 1D barcodes with camera phones that work only as a CCD scanner. Continuous network connectivity also must be guaranteed to use the system feasibly. However, considering the global proliferation of 1D barcode and its systems, the use of 1D barcode as a ubiquitous computing tool is worthy of further investigation.

2.4.2   RFID, NFC and barcode

Barcode technology has been compared with another technology with growing popularity, namely RFID technology [13]. Radiofrequency identification (RFID) is a system that allows contact-free automatic identification. Data stored in an RFID tag are transmitted via radio waves, and read by a reader. Radiofrequency identification (RFID), smart cards and magnetic stripes have often been used where there is a need for changes in the information encoded.

Recently, one particular type of RFID technology called near field communication (NFC) has attracted massive interest. Near field communication is a standards-based short-range wireless-connectivity technology. It enables two-way communications with peer electronic devices operating in the 13.56 MHz range over a short distance (i.e. up to 20 cm). The NFC technology is different from the other two technologies (i.e. barcode and RFID) in that it can work actively, owing to its capability of two-way communication.

A NFC chip can work not only as a tag that carries particular information but also as a reader to interpret the information provided by other devices nearby, allowing the technology to offer different types of application from those offered by RFID. For example, by placing a mobile phone with a NFC chip over a movie poster to which another NFC chip is attached, a user can collect the information about the movie (e.g. time and the venue). The user can then purchase a ticket, by waving the mobile phone in front of the ticket machine sending information about the movie.

The RFID, NFC and barcode technologies have been in the spotlight as ubiquitous computing tools, offering similar applications. However, each has its advantages and disadvantages. Table 2.4 presents for comparison the characteristics of each technology. Reading barcodes has restrictions, requiring a clear line of sight; in contrast, RFID technology can read tags even when they are out of sight. Furthermore, more than one RFID tag can be read at once. The long-lasting nature of an RFID tag is also becoming valued. However, advantages from one perspective can be disadvantages when viewed from another.

Unlike barcode, RFID technology does not require a clear line of sight with which to interact. Malicious users are able to access information contained in the RFID without being recognised. The indestructability of RFID can also be seen as a security or privacy threat. For example, RFID tags that are no longer in use might be kept somewhere, containing a product ID along with the credit card details of a person who purchased it. Conversely, once data are encoded in the format of a barcode, the data inside cannot be changed without visual alteration. Furthermore, physical access to the barcode symbol is necessary.

The operation cost of an RFID system has also been an obstacle for widespread adoption of this technology. Nonetheless, with the remarkable progress in microelectronics and low-power semiconductor technologies, inexpensive RFID tags are becoming a reality. In the near future, the price of RFID may fall below a critical threshold [22]. However, in comparison with that of RFID technology, the cost of using barcode is negligible.

Table 2.4. Comparison of the characteristics of barcode, RFID and NFC technology

	1D barcode	2D barcode	RFID	NFC
Data accuracy	◯	◯	◯	◯
Data capacity	×	◯	◯	◯
Reading distance	◯	◯	◯	×
Reading capability	×	×	◯	◯
Human readability	◯	×	×	×
Operation cost	◯	◯	×	×
Security	△	◯	×	×
Two-way interaction	×	×	×	◯

◯, high; △, medium; ×, low.

No one technology satisfies all the requirements for use as a ubiquitous computing tool. It is expected that all these technologies will make further progress but still complement one another.

2.4.3 Barcode technology and ecology

One of the most remarkable aspects of barcode technology is its environmental friendliness. In general, barcode technology uses only ordinary paper and ink. Some applications use barcodes that are digitally displayed on a medium such as a camera phone display or on a personal computer (PC) screen. Such barcodes are re-writable and can be created, saved, exchanged and/or used as much as required with no waste and no additional cost.

At present, barcode use is limited to the applications for mobile devices such as camera phones and personal digital assistants (PDAs). However, it is possible to develop applications that use a similar medium for industrial use, which may rapidly escalate the demand for barcode technology.

Recently, the direct-marking techniques of barcodes have also attracted the attention of certain interest groups, e.g. the Air Transport Association (ATA), the National Aeronautics and Space Administration (NASA) and the Automotive Industry Action Group (AIAG), since they neither require additional cost nor produce waste. A barcode directly marked on the surface of a material can be used where printed labels would not adhere or would be destroyed. This improves the trackability of products, allowing manufacturers to manage a product throughout its life, from its production through to disposal. A system that can control and manage the life of each product is essential for enhancing product reuse or recycling. Hence, the adoption of direct-marking techniques is related to the environmental friendliness of barcode technology.

2.5 Summary

The initial reasons for the use of keyless data-entry and automatic-identification systems were to minimise errors caused by human mistakes during data entry and to process data efficiently. In response to such a demand, different technologies and their systems have been developed, including optical character recognition (OCR), RFID, magnetic-stripe, NFC and barcode technology. One of the most distinctive advantages of barcode use is its inexpensive operation. Once its fast and robust operation, usability and effectiveness were demonstrated in the USA, 1D barcode technologies and their systems spread first to European countries and then all over the world. The last few decades have witnessed their huge success worldwide, becoming a necessity of daily life.

In addition to 1D barcodes, 2D barcodes and 3D barcodes now exist. They were developed to respond to needs in various fields. Instead of competing with each other, 1D, 2D and 3D barcode systems have co-existed, each opening up its own market. One of the distinctive differences between 1D barcode and 2D barcode is in their data capacity. One-dimensional barcode can encode only a limited number of digits, alphanumeric characters and special characters, whereas its 2D counterpart has a much higher data capacity. Thus 1D barcode is merely a key to a database, whereas 2D barcode can work as a portable database.

There is another difference between 1D barcode and 2D barcode that is worthy of special mention. A number of 2D barcode symbologies have been developed independently in response to the demand from certain industries. Meanwhile, 1D barcode's code and symbol selection has been based on the necessity for all interest groups to use the same symbol structure and for each code to be uniquely identified. That is, 1D barcode has been developed with the intention of global use. Hence, despite its shortcoming in data capacity, 1D barcode has been widely accepted throughout the world and presumably this trend will continue.

Recently, mobile phones equipped with CCD cameras have gained great popularity. The integration of camera phone and 2D barcode technology has opened a door for the development of novel mobile applications. In such applications, 2D barcodes work as a ubiquitous computing tool, bridging the physical and digital worlds. The RFID and NFC technologies have also attracted attention as suitable tools for ubiquitous computing. Hence, the RFID, NFC and 2D barcode technologies are often compared with each other. As with the different types of barcode technologies, it is expected that they will co-exist.

In addition to the great potential for application development, another meaningful outcome of the integration of camera phone and 2D barcode technologies is that 2D barcode symbols can be recorded on re-writable media such as mobile phone displays and stored in hardware memory. This enables paperless barcode applications, making them environmentally friendly.

A variety of 2D barcode symbologies for mobile application have been developed and the trend is still progressing. In Chapter 3, these 2D barcodes will be described in greater detail.

3 Two-dimensional barcode for mobile phones

Since their first appearance, 2D barcodes have evolved with additional features including security, error detection and correction capability and the ability to encode different languages. The ability to encode a robust portable data file has made 2D symbologies attractive regardless of their extra space requirements. Accordingly, industries and organisations in various fields have adopted applications that use 2D barcodes within their systems. These range from medicine labelling for the healthcare industry and the marking of small items such as integrated circuit boards for IT vendors through to the secure transmission of battle-field plans in the defence industry. However, despite its potential for wider use, this technology has been adopted by only a few industries and organisations.

Mobile devices such as mobile phones and personal digital assistants (PDAs) are another notable technology that has evolved during the last decade. With their current sophisticated functions these devices can provide an entertainment centre, performing the role of a game console, a jukebox, a digital camera and a social organiser. As more advanced features and applications are developed, they should become less expensive and more commonly used. In fact, PDAs and mobile phones are now virtually universally used in industrialised countries across different age groups. Furthermore, they are always available, providing seamless and ubiquitous connectivity. As the research firm Gartner has predicted [19], this trend is expected to continue for some years.

Two-dimensional barcodes and mobile phones have a lot in common. They have become smaller, yet smarter, in the last decade and are still being improved. Both technologies are portable, allowing them to operate anywhere; 2D barcodes ubiquitously provide the source of an action, which is interpreted by the built-in camera of a mobile phone and implemented by the software installed in the phone. The connectivity inherent in a mobile phone allows such local actions or events to become global, providing uninterrupted access to different locations in both the physical and digital worlds. Users can cost-effectively enjoy a variety of applications based on connections established via the integration of 2D barcode and camera phone technology, regardless of time and location.

Despite their capability, both 2D barcodes and mobile phones can operate inexpensively. The cost of using barcode technology is negligible. The recent trend is that computing devices are becoming smaller, less expensive and more abundant [23], and the mobile phone is no exception. In addition to the technological aspects, the economical advantage indicates the positive contribution of these two portable technologies towards the realisation of ubiquitous computing. As a result, numerous mobile applications

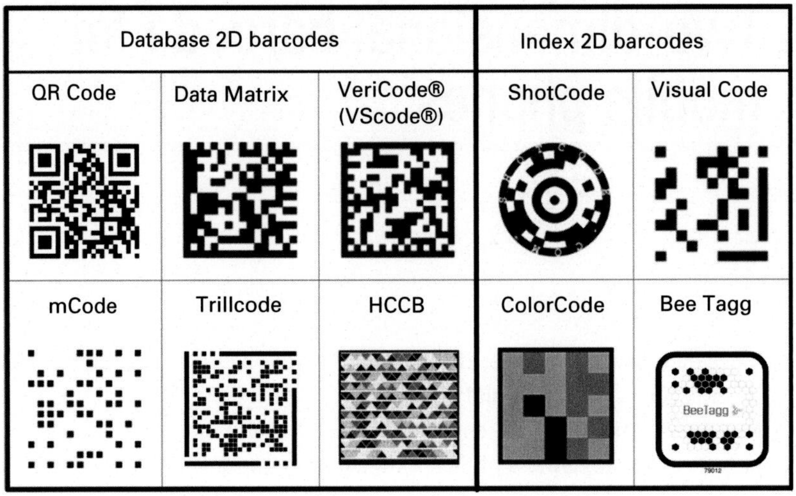

Database 2D barcodes			Index 2D barcodes	
QR Code	Data Matrix	VeriCode® (VScode®)	ShotCode	Visual Code
mCode	Trillcode	HCCB	ColorCode	Bee Tagg

Fig. 3.1   Examples of 2D barcodes for mobile applications. The derivatives of QR Code and Data Matrix are introduced in the Sections 3.1 and 3.2. Note that both HCCB and ColorCode use colour.

have been devised using existing 2D barcodes that were initially developed for industrial use. Some researchers and developers have invented new 2D barcodes, targeting camera phone users. Figure 3.1 presents examples of popular 2D barcodes currently in use.

The 2D barcodes used for mobile applications fall into two categories; those in the first category were invented initially to improve capacity for encoding raw data. Furthermore, various supplementary features can be added to increase the robustness of 2D barcode, making the most of their improved data capacity. These features include an error detection and correction capability, multilingual encoding, a structured append feature and a data compaction scheme. The resultant 2D barcode symbol can work as a robust, portable, 'data file' or 'database'. In this book, we refer to such symbols as database 2D barcode.

The barcodes in the second category were developed for camera phone applications, taking into account the reading limitations of the built-in camera of mobile phones. They also differ greatly from the former 2D barcodes in terms of data capacity, as their focus is more on providing robust and reliable barcode reading. Each barcode basically works as an index linking the physical and digital worlds. Continuous network connectivity must be guaranteed for them to work ubiquitously. They will be referred to as index 2D barcode in this book. Index 2D barcodes function in a similar way to their 1D counterparts.

The three database 2D barcodes in the upper row in Figure 3.1, namely, QR Code, Data Matrix and VeriCode, were originally developed for industrial use, each having been optimised for specific applications. However, technologies often thrive when people begin using them for purposes other than those intended by their inventors. In fact, QR Code and Data Matrix have become two of the most widely accepted 2D barcodes for mobile applications. On the other hand, mCode and Trillcode, in the lower row, are database 2D barcodes specifically designed to meet the needs of emerging camera phone applications. High Capacity Color Barcode (HCCB), developed within Microsoft Research, is a database 2D barcode that uses colour to increase the data capacity. This code was invented with the aim of targeting both industrial use and mobile applications.

Of the four index 2D barcodes, the two in the upper row of Figure 3.1 (i.e. ShotCode and Visual Code) were invented specifically for human–computer interaction, whereas the 2D barcodes in the lower row, namely, ColorCode and BeeTagg, were designed to help consumers establish easy, error-free, Web links. ShotCode, originally called SpotCode, was acquired by a Swedish company called OP3 and given its present name and new tasks, which are similar to the tasks of ColorCode and BeeTagg. ColorCode is the first 2D barcode that utilises colour to encode data for mobile applications.

A full description of each 2D barcode will be given in the following sections.

3.1   QR Code

3.1.1   Overview

Quick Response (QR) Code[1] was developed by Denso Wave (a division of Denso Co. at the time) in 1994 [24]. This code is a two-dimensional matrix symbology that has position-detection patterns at three corners. As the name suggests, it was initially designed for ultra-high-speed and omnidirectional reading [25]. Thus, QR Code was developed to improve the reading speed of complex-structured 2D barcodes. It is also known for its capability of directly encoding the Japanese and Sino–Japanese Kanji–Kana character sets. Other QR Code features are massive data capacity, high data density and selectable levels for error correction ability [1]. Recently QR Code derivatives have been developed; these include Mobile Code and MS-Code.

Quick Response Code has been approved as a standard both in Japan and internationally. These standards are as follows.

(i) Automatic Identification Manufacturers (AIM) International standard ISS – QR Code;
(ii) Japanese Electronic Industry Development Association standard JEIDA-55;
(iii) Japanese Industrial Standard JIS X 0510; and
(iv) International Organization for Standardization (ISO) standard ISO/IEC 18004.

In August 2002 in Japan, J-SH09 (the manufacturer was Sharp and the carrier was J-Phone[2]) was released as the first mobile phone that had a reader for Japanese Article Number (JAN) code (1D barcode) and QR Code (2D barcode) as one of its functions [26]. Many other Japanese phone companies such as DoCoMo, Vodafone (now, Softbank Mobile Co.) and KDDI au also began offering mobile phones with the ability to read QR Code [26].

---

[1] The material on QR Code and Micro QR Code was referenced from the Information Technology Automatic Identification and Data Capture Techniques QR Code 2005 Barcode Symbology standard ISO/IEC 18004 and from the standard ISO/IEC SC 31 of the National Body of Japan, unless otherwise indicated.
[2] Now Softbank Mobile Co.

3.1.2   QR Code symbol structure

Quick Response Code consists of seven elements, namely, a finder pattern, a timing pattern, an alignment pattern and the quiet zone (making up the function patterns) and the format information, separator and data areas (making up the encode area). Figure 3.2 shows the code structure of QR Code. The finder pattern is usually called the position-detection pattern and the separate parts of it are located at the three corners of the symbol. The timing pattern consists of broken-line borders placed between the blocks of the finder pattern. The blocks of the alignment pattern are assigned within the code area in fixed positions and each block has an isolated cell at its centre. Each symbol is surrounded by a four-module-wide quiet zone (two-module wide for Micro QR Code). The function patterns are used to define first the accurate position of the symbol, then the symbol size, its orientation and the whole aspect of the symbol. The data field includes Reed–Solomon codes as well as raw data. The format information includes the symbol version, the error correction level and the mask number.

Position-detection pattern

The position-detection (finder) pattern blocks at the three corners of a symbol are one of the most distinctive features of QR Code. When scanned, these pattern blocks are the first to be detected by the reader, which then locates the position of the code at ultra-speed. The ratios of the black and white lengths on a line that passes through the centre of the finder pattern (i.e. B : W : B : W : B) are 1 : 1 : 3 : 1 : 1 at any angle (see the right-hand panel of Figure 3.2). This unique set of ratios enables fast detection of the three finder pattern blocks in the symbol. Once the position of the symbol is located, the code size L, tilt θ and orientation are calculated from the positions of the three finder patterns (see Figure 3.3). This allows the symbol to be read omnidirectionally.

Timing pattern

The timing pattern consists of vertical and horizontal broken-line borders placed between the finder pattern blocks (see Figure 3.4). These borders are used to calculate the centroid of each cell and modify it when symbol distortion and/or changes in cell pitch are found [13].

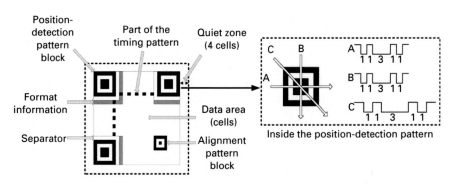

Fig. 3.2   Quick Response (QR) Code symbol structure.

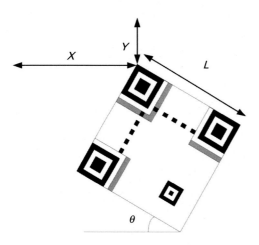

Fig. 3.3   Orientation of a QR Code symbol (adapted from [13]).

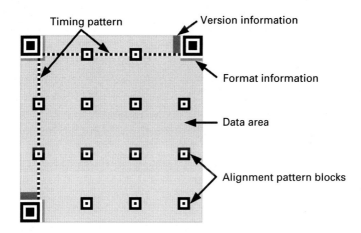

Fig. 3.4   Timing pattern, alignment pattern and data area.

Alignment pattern

The alignment pattern (Figure 3.4) enables the correction of any local distortion [25]. This is achieved by determining the centre coordinates of each alignment pattern block and then adjusting the centroids of the data cells accordingly. The isolated black cell within each alignment pattern block enables fast calculation of the centre coordinate of the block.

Format information

The format information indicates the symbol version, the error correction level (see Section 3.1.3) and the mask pattern (see Section 3.1.4) used for the symbol. Therefore, this area is the first place to be read in the decoding process.

Data area

The data area is where the original data (i.e. the raw data) and the Reed–Solomon code are encoded (see Figures 3.2 and 3.4). The Reed–Solomon code is a mathematical error correction method originally developed for artificial satellites and planetary probes as a measure against communication noise. It is capable of making a correction at the byte level and is suitable for concentrated burst errors [24].

Quiet zone

The QR Code requires a four-module-wide quiet zone (margin) at all sides of the symbol (there is a two-module-wide margin for Micro QR Code). This area allows the symbol to be distinguished from its background, which leads to accurate high-speed reading.

3.1.3   Symbol description

QR Code models

Three different models of QR Code exist: Model 1, Model 2 and Micro QR Code (see Figure 3.5).

The QR Code Model 1 is the original QR Code; the other two models are derivatives of Model 1 [27]. There are 14 different versions of QR Code Model 1, Version 1 to Version 14, and they have been approved as AIM international standards (ISS – QR Code). In order to improve its robustness against distortion, alignment patterns have been added to QR Code Model 1 [27]. The outcome is called QR Code Model 2 and has 40 versions, Version 1 to Version 40; all the versions have been approved as AIM international standards. Micro QR Code is a small-sized QR Code that fits applications where less space and data are required. The minimum symbol size of Micro QR Code is as small as 11×11 cells, in contrast with ordinary QR Codes whose minimum size is nearly twice as large, 21×21 cells [27].

Symbol version

Small black and white square cells make up QR Code, and they are called modules. This code has 40 symbol versions, ranging from Version 1 (21×21 modules) to Version 40 (177×177 modules). Each version has a different module configuration or number of

QR Code Model 1   QR Code Model 2   Micro QR Code

Alignment pattern

Fig. 3.5   QR Code models. Note that an alignment pattern is present only in Model 2 and is indicated by the grey arrow.

Distorted code   Dirty code   Torn code

Fig. 3.6   Examples of damaged QR Code symbols.

modules. Four modules are added on each side of the symbol as the version number increases. The maximum data capacity for each QR Code symbol version is determined in accordance with the amount of data needed, the character type and the error correction level. Figure E.1 and Tables E.1 to E.4 in Appendix E show the QR Code symbol versions and the maximum data capacity for each symbol version.

Error correction

Quick Response Code has an error correction capability to restore the original data if the barcode becomes dirty or damaged. A maximum of 30% of codewords can be restored even if a QR Code symbol is damaged, as presented in Figure 3.6.

The error correction is implemented by adding Reed–Solomon code to the original data. As presented in Table 3.1,[3] four different error correction levels are available: L (approx. 7%); M (approx. 15%); Q (approx. 25%); and H (approx. 30%). Users have a choice according to their need and the operating environment. A higher error correction level improves the error correction capability but also increases the amount of data to be encoded, which results in a larger symbol size. Level M (15%) is most frequently selected.

The error correction level should be chosen according to how much data needs to be corrected. For example, when 50 out of 100 codewords of QR Code need to be corrected, 100 codewords of Reed–Solomon code are required, as the amount of Reed–Solomon code required is twice the number of codewords to be corrected. Hence, the total number of codewords becomes 200. That is, 50 codewords out of 200 can be corrected. This brings a 25% error correction rate in terms of the total number of codewords and thus corresponds to QR Code error correction level Q.

Symbol size

The actual size of a QR Code symbol is determined by the millimetre size of the module [24]. As the module size increases, the code becomes more stable and more easily read with the scanner. It is recommended, therefore, that QR Code symbols be printed as large as possible within the available printing area.

---

[3] A codeword is a unit that constructs the data area. One codeword in QR Code is equal to eight bits.

Table 3.1. QR Code total data
restoration rate for codewords

Level L	Approx. 7%
Level M	Approx. 15%
Level Q	Approx. 25%
Level H	Approx. 30%

Table 3.2. QR Code maximum data capacity and data translation ratio

Maximum data capacity	Numeric	Max. 7089 characters
	Alphanumeric	Max. 4296 characters
	Binary (8 bits)	Max. 2953 bytes
	Kanji, full-width Kana	Max. 1817 characters
Data translation ratio (number of cells	Numeric	3.3 cells/character
needed to encode a character)	Alphanumeric	5.5 cells/character
	Binary (8 bits)	8 cells/character
	Kanji, full-width Kana	13 cells/character

Encodable character set

Quick Response Code is capable of encoding all types of data, such as numeric and alphanumeric, Japanese (Kanji, Kana and Hiragana) symbols, binary, control codes and image data [27]. The maximum data capacity and data translation ratio are presented in Table 3.2.

Since QR Code was developed in Japan, provision was made to encode Japanese Industrial Standards (JIS) Level 1 and Level 2 Kanji character sets. One full-width Kana or Kanji character can be efficiently encoded in 13 bits, which allows QR Code to contain 20% more data than other 2D symbologies when encoding Japanese characters.

3.1.4   Advantageous features

Flexibility in symbol format

A colour reversal function allows the cells in QR Code symbols to be printed light-on-dark or dark-on-light. Furthermore, QR Code can encode and decode round cells. This feature is especially advantageous for direct part marking since 2D barcodes may be marked on materials that are not necessarily a light colour. In cases where a laser marking technique is used, the cells may appear round. When QR Code is marked on a transparent material such as glass, the symbol can be decoded from the back [13] (see Figure 3.7).

Masking

Quick Response Code masking is a technique used to enable easy and quick barcode reading. The masking technique allocates black and white dots evenly and helps prevent

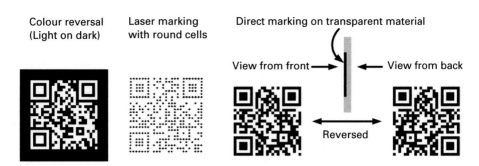

Colour reversal
(Light on dark)

Laser marking
with round cells

Direct marking on transparent material

View from front → ← View from back

Reversed

Fig. 3.7   Different types of direct marking (adapted from [13]).

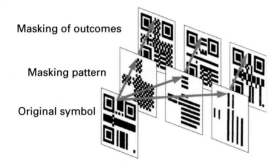

Masking of outcomes

Masking pattern

Original symbol

Fig. 3.8   Masking operation (adapted from [13]). One masking outcome is chosen out of the eight possibilities, only three of which are shown.

pattern duplication. It also helps to avoid the data area having the same black and white pattern as that used in the finder pattern (i.e. $1:1:3:1:1$) since that could confuse the reading program and slow its performance. Eight masking patterns exist, and the encoders evaluate each pattern to select the best (see Section 5.2.2). The evaluation is conducted in the following manner. After applying each masking pattern to the symbol, if there are still lines or blocks that consist of only one colour (i.e. they are all black or all white), the suitability score for the masking pattern will be reduced. If finder-pattern-like lines or blocks are observed, the suitability score will be reduced further. The masking pattern that earns the highest score is selected [13], [27]. The masking operation is shown in Figure 3.8.

Structured append feature

The structured append function enhances a QR Code's scalability by dividing the code into up to 16 data areas. In this way, a large QR Code symbol can be broken into smaller ones (see Figure 5.7), so that the symbol cell size and data capacity can be manipulated to manage space restriction and the limitations of camera phone applications. A reader can then reconstruct the information that is stored in the multiple QR Code symbols as single data symbols. A detailed description of the QR Code structured append will be given in Chapter 5.

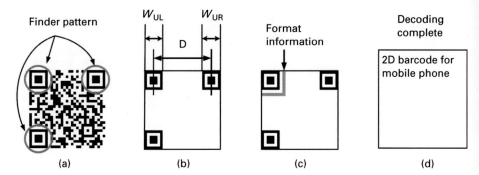

Fig. 3.9   Decoding process of QR Code.

3.1.5   Scanning and decoding

Charge coupled device (CCD) scanners are used for capturing the 2D barcode symbol image. The structure of the CCD imager for 2D barcode is presented in Figure 2.17. A built-in CCD camera in a mobile phone operates in the same way, although there might be a difference in the quality of the functions. Figure 3.9 shows the decoding process for QR Code, once the symbol image is captured.

(i) The three finder patterns are located and the centre of each pattern is calculated (Figure 3.9(a)).

(ii) The module size of the symbol is determined by measuring W_{UL} and W_{UR} and the symbol size is determined by calculating the size of D (Figure 3.9(b)).

(iii) The format information is decoded and the error correction level and masking pattern to be used are defined (Figure 3.9(c)).

(iv) The whole data block is detected and then the codewords for error correction are removed. The raw data are decoded according to the defined error correction level and masking pattern. As a result of decoding, the encoded text, '2D barcode for mobile phone' is displayed (Figure 3.9(d)).

3.1.6   Applications

There has been a remarkable increase in the number of camera phone users who have shown an interest in QR Code applications. This trend is especially true for younger users.

Users can create QR Codes using free barcode generators. They can then be exchanged over the Internet, saved on mobile phones or printed. Furthermore, some users can even develop their own applications. An example is the QR Code Blog, which is written in QR Code symbols.

The medium most often used for displaying QR Code is a newspaper or magazine and the data type most often decoded by camera phones is a uniform resource locator (URL) followed by an email address. Thus, the most popular QR Code application is the linking of camera phones to the Internet, where users have uninterrupted access to on-line shopping and real-time information on public transport and events. Being one of the database 2D barcodes, QR Code can also be used when no network connectivity is available. For example, by simply capturing a QR Code symbol printed on a business card, users can store in their phone address books the contact information encoded in the symbol.

Quick Response Code is also used for portable files such as e-tickets and e-coupons. Furthermore, QR Code applications that are relevant to a consumer's everyday life have become popular recently. For example, in responding to growing concerns over produce safety, agricultural organisations have implemented food-history information services as part of projects to improve produce quality, and also assure consumers' food safety, by providing the produce detail. Quick Response Code works as a medium for tracking, a portable file that carries produce information and a tag to provide a link between producers and consumers. By capturing the QR Code on the package, consumers can access information on the produce, including the supplier's contact details, harvest and shipment dates and the fertilisers and agricultural chemicals used.

3.1.7   Micro QR Code

Micro QR Code is a small-sized QR Code that fits applications where less space and data are required. Figure 3.10 shows the symbol structure of Micro QR Code.

The main structural difference between QR Code and Micro QR Code is the number of position-detection patterns and the space required for the quiet zone. Each Micro QR Code symbol uses only one position-detection pattern whereas three position-detection patterns are used for a QR Code symbol. Although a QR Code symbol requires a four-module-wide quiet zone at all sides of the symbol, a Micro QR Code is surrounded by an only two-module-wide quiet zone.

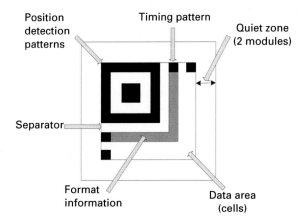

Fig. 3.10   Micro QR Code symbol structure.

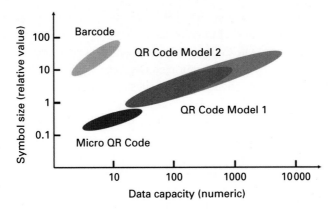

Fig. 3.11   Barcode relative symbol sizes (adapted from ISO/IEC SC 31, National Body of Japan, 2002).

The printed labels for circuit boards and electronics parts are examples of Micro QR Code usage. The efficiency of data encoding is increased with the use of only one position-detection pattern. Figure 3.11 gives a symbol-size comparison of linear barcode, Micro QR Code, and QR Code, Models 1 and 2.

3.2   Data Matrix

3.2.1   Overview

Data Matrix[4] is a two-dimensional matrix symbology invented by International Data Matrix Inc. in 1987 [28]. At present, two main subsets of the Data Matrix code exist, ECC 000–140 and ECC 200 (ECC stands for 'Error Correcting and Checking'). The former includes ECC 000, ECC 50, ECC 080, ECC 100 and ECC 140, the original versions of the Data Matrix code. In 1995, a new version of the Data Matrix code was developed by Data Matrix Inc. to improve the data capacity and reading robustness of Data Matrix ECC 000–140. The resultant symbology is referenced as ECC 200.

Data Matrix symbology adopts two types of error correction algorithm, depending on the ECC level employed. The ECC levels 000 to 140, which offer five different error correction levels,[5] use convolutional code error correction. However, the commonly used ECC 200 level uses Reed–Solomon error correction. The name of each version was derived from the ECC level. The correction level of ECC 200 is determined by the symbol size.

The symbol sizes of Data Matrix ECC 000–140 range from a minimum, 9 × 9 cells, to a maximum, 49 × 49 cells. Unlike ECC 200, the number of cells for ECC 000–140 is always odd. A two-dimensional imaging device such as a CCD camera is necessary

---

[4] The material on Data Matrix code was referenced from Information technology – International symbology specification – Data Matrix (ISO/IEC 16022), unless otherwise indicated.

[5] ECC 000 (0%), ECC 50 (2.8%), ECC 080 (5.5%), ECC 100 (12.6%), ECC 140 (25%). ECC 000 to ECC 140 are still available.

to scan the symbol. Although Data Matrix ECC 000–140 symbols are still available and valid, their usage is limited owing to their inferiority to ECC 200. For this book, the focus is mostly on Data Matrix ECC 200.

Data Matrix supports industrial-standard escape sequences to define international code pages and special encoding schemes and is globally used for small-item marking applications employing a wide variety of printing and marking technologies. In 1996, Data Matrix ECC 200 was approved as the American National Standards Institute ANSI/AIM BC11 International Symbology Specification (ISS – Data Matrix) standard. Data Matrix ECC 200 was also accepted as an ISO international standard (ISO/IEC 16022) in 2000. Data Matrix symbology has been standardised domestically as well as internationally. Organisations and industries which have adopted Data Matrix include the Automotive Industry Action Group (AIAG), the Electronic Industries Alliance (EIA) and National Aeronautics and Space Administration (NASA).

Just as in the case of QR Code, the derivatives of Data Matrix symbology have often been used for mobile applications.

3.2.2   Data Matrix symbol structure

Each Data Matrix symbol contains dark and light square data modules, which are comparable with the cells for other 2D barcodes. It has a finder pattern of two solid lines and two broken lines forming its perimeter. The symbol is surrounded on all four sides by a quiet zone border. Figure 3.12 shows the symbol structure of Data Matrix.

The L-shaped solid border is used primarily to define the physical size of the symbol and its orientation and distortion, whereas the broken borders at the opposite corner define the symbol's cell structure. The latter borders also assist in determining the physical size and distortion of the symbol [4].

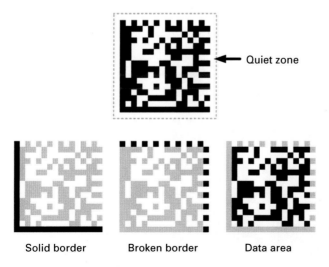

← Quiet zone

Solid border   Broken border   Data area

Fig. 3.12   Data Matrix symbol structure. A particular feature is highlighted in each diagram.

3.2.3   Symbol description

Symbol shape

Data Matrix symbols can be either square or rectangular; both symbols in Figure 3.13 contain the same amount of data.

Symbol size

Unlike Data Matrix ECC 000–140, Data Matrix ECC 200 symbols have an even number of modules on each side. The maximum size of each block is 24 × 24 modules, which helps to prevent symbol distortion. For the larger symbol sizes in ECC 000–140, the first read rate drops dramatically due to symbol distortion. A block structure was added to ECC 200 to remedy the shortcoming of ECC 000–140. Thus, a symbol is divided into blocks once the size exceeds 24 × 24 modules. For example, a symbol that contains 26 × 26 modules is divided into four blocks (see Figure 3.14).

Data Matrix data capacity

As indicated in Table 3.3, the minimum data capacity of an ECC 200 symbol is six numeric digits, three alphanumeric characters or 1 byte in a symbol 10 modules square,

Square symbol   Rectangle symbol

Fig. 3.13   Data Matrix symbol shapes.

Fig. 3.14   Data Matrix symbol divided into four blocks.

Table 3.3. (a) Minimum and (b) maximum data capacity for ECC 200 symbols

(a) Size	Data capacity	
10 × 10 modules	numeric	6 characters
	alphanumeric	3 characters
	8-bit byte	1 byte

(b) Size	Data capacity	
144 × 144 modules	numeric	3116 characters
	alphanumeric	2335 characters
	8-bit byte	1556 byte

whereas the maximum data capacity is 3116 numeric digits, 2335 alphanumeric characters or 1556 bytes in a symbol 144 modules square. That is, the symbol is scalable between 1 mm^2 to 14 inch^2 (i.e. 355.6 mm^2) [15].

Error correction

Data Matrix ECC 200 uses the Reed–Solomon error correction algorithm. As opposed to ECC 000–140, whose error correction levels can be chosen as required by certain applications, the error correction level of ECC 200 is automatically determined in accordance with the symbol size (see Appendix F). Hence, users cannot select the error correction level.

A codeword is the unit that constructs the data area of a Data Matrix symbol. The error correction capability of Data Matrix is measured by the restoration percentage for the entire data. The data restoration rate for misread symbols and that for symbols that could not be read at the first attempt are different. The former varies from 14.2% to 25.0%, whereas it is much higher for the latter, ranging from 27.1% to 38.9% [13].

Encodable character set

Two types of character set can be encoded in Data Matrix symbols: the 128 characters conforming to ISO 646 or user-defined extended character sets of 256 characters. The latter option encodes all the data byte by byte.

3.2.4 Advantageous features

One of the noteworthy Data Matrix features is its small symbol size: Data Matrix uses the smallest symbol (3.3 × 3.3 mm) of the three database 2D barcodes (QR Code, Data Matrix and VeriCode) in encoding identical data (14 digits) in the same cell size (0.25 mm) and with the same level of error correction (10%–15%) [29]. Other features that can enhance Data Matrix's reading robustness include colour reversal, data compaction, code pages and a structured append function.

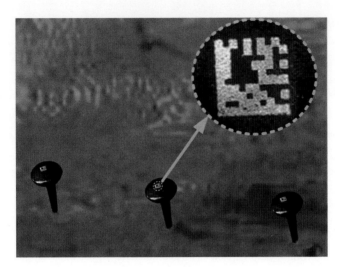

Fig. 3.15   Light-on-dark Data Matrix symbol on the head of a straight pin.

Colour reversal

As in the case of QR Code, the modules in Data Matrix symbols can be printed light-on-dark or dark-on-light. This feature is useful for direct part marking, where 2D barcodes may need to be marked on materials that are not necessarily a light colour [13]. Figure 3.15 demonstrates how small a Data Matrix symbol can be, together with its colour reversal feature.

Data compaction

Data compaction is a scheme condensing beyond eight bits per character to increase the data capacity of 2D barcode symbols [4]. This is done prior to the encoding of data in a Data Matrix symbol format.

Code pages

Code pages provide an optional feature in ECC 200 which enables the encoding of characters from other parts of ISO 8859 (e.g. Arabic, Cyrillic, Greek, Hebrew) as well as other data interpretations or industry-specific requirements. Up to 256 different code pages can be specified [4].

Structured append

This feature allows a large amount of data to be represented logically and continuously in up to 16 Data Matrix symbols. The original data can be correctly reconstructed regardless of the symbols' scanning order [4]. Three additional bytes are required for this function to work. The first four bits of the first byte identify the position of the particular symbol in the sequence (see Figure 3.16). The last four bits identify the total number of symbols in the sequence. The second and third bytes are used as a file identifier with possible values between 1 and 254 (thus allowing up to $254 \times 254 = 64\,516$ identifiers) [28].

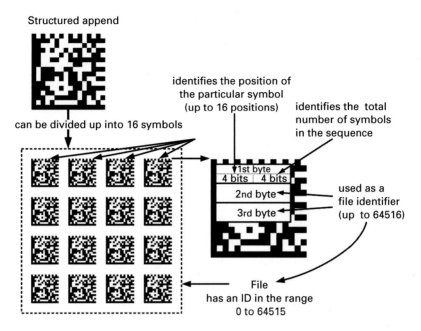

Fig. 3.16   Data Matrix structured append operation.

3.2.5   Scanning and decoding

Using a CCD video camera or a CCD scanner, Data Matrix symbols can be read:

   (i) omnidirectionally;

  (ii) with a contrast ratio as low as 20%;

 (iii) even if a substantial portion of the mark is missing or obscured;

 (iv) at distances ranging from close contact to 36 inches (91.44 cm) away; and

  (v) at a rate of five symbols per second [15].

Data Matrix symbol decoding is a two-step process: image recognition and retrieval of the original data. As an example, the Data Matrix symbol decoding process provided by the Semacode organisation is presented below.

Image recognition and retrieval of the original data

Once the image has been captured by a phone's camera subsystem, the reading software:

   (i) locates the Data Matrix symbol inside the image;

  (ii) corrects the image distortion; and

 (iii) acquires the data from the image.

Several techniques from the field of image recognition are applied to detect the edges of the symbol, to discard objects that are not part of the symbol and to determine the logical geometry of the symbol. Once the image is detected, image distortion including any rotation, perspective error or fuzzy edges must be corrected. For mobile phone use, these operations are implemented within the confines of a limited memory and a relatively slow central processing unit (CPU).

The raw data result is the value of black and white (i.e. binary '1' or '0') for each module in the symbol on a two-dimensional matrix grid [30]. The reading software then decodes the data by applying Reed–Solomon error correction and decoding. The result of this stage if successful is a decoded message, in other words, a string of characters.

3.2.6   Data Matrix derivatives and mobile applications

Some companies and organisations have adopted the Data Matrix format to implement their own mobile applications. Examples of such symbologies are Semacode and UpCode. Semacode Co. is one of the pioneer companies that utilised the existing 2D barcode (i.e. the Data Matrix format in this case) for mobile applications. Semacode is used to encode plain-text URLs and implement them as physical hyperlinks that allow a mobile phone's browser to access the designated Web page [30]. It is also utilised in applications such as business cards, conference badges and the live delivery of urban geographical information. UpCode, from the Finnish company of the same name, provides applications that allow users to interact with contents on the Web, by means of a URL link established via a camera phone. The applications include business cards, payment and customer relationship management (CRM) solutions.

3.3   VeriCode

3.3.1   Overview

VeriCode,[6] from Veritec Inc., uses unique symbol-encryption techniques to provide additional security for mobile applications. A distinctive feature of VSCode, a derivative of VeriCode, is its high data capacity; this allows the code to store biometric data for access control applications.

The amount of information within a VeriCode symbol is user selectable, which determines the density of the symbol. This helps solve the resolution problem of camera phones. VeriCode also contains a high percentage of error detection and correction (EDAC) capability (15%–25%) to ensure its reading robustness. Its ability to mark directly most materials, including metal, glass and plastic, is a key strength [4].

3.3.2   VeriCode and VSCode symbol structure

VeriCode and VSCode consist of a solid border frame, a data field and an invisible border (i.e. quiet zone). The solid border frame encloses the internal data field of single cell units representing binary data. The quiet-zone border must be at least one data cell wide

---

[6] The material on VeriCode and VSCode was referenced from [31], unless otherwise indicated.

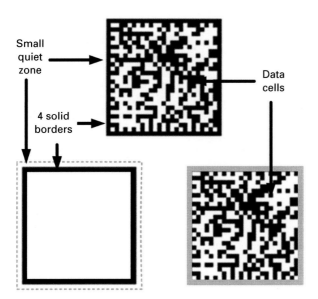

Fig. 3.17 Symbol structure of VeriCode and VSCode.

and is critical to separating the data field from surrounding noise [31]. Figure 3.17 shows the symbol structure of VeriCode and VSCode.

3.3.3 VSCode

Each VSCode symbol can store more than 4000 bytes of information, making VSCode a secure container as well as a portable data storage unit [31]. This portable data storage capability is ideal for storing personal, financial and biometric information. The data stored within a VSCode symbol can be obtained instantly, without accessing a central database, and can be securely protected with an individualised biometric encryption key.

A VSCode symbol shares many features of VeriCode but has a much larger data capacity, storing up to 4151 bytes of data, whereas the maximum data capacity of VeriCode is 500 bytes. This is the highest-capacity black and white 2D barcode. The description of VeriCode provided in the following section applies to both VeriCode and VSCode.

3.3.4 Symbol description

Symbol shape and size

VeriCode symbols can be either square or rectangular in shape; both symbols in Figure 3.18 contain the same amount of data.

The size of a VeriCode symbol is variable and scalable to fit the available marking space. Therefore, the data density is not a function of the symbol size. This allows users to select the size of VeriCode symbol on the basis of their space requirements.

Square symbol

Rectangle symbol

Fig. 3.18   VeriCode symbol shapes.

Encodable character set

VeriCode is capable of containing any information that can be converted into a stream of binary data; any language or character set can be encoded. A Vericode symbol can be used as a stand-alone data field or as a key to a database file.

Error correction

VeriCode contains high-percentage EDAC capability, ranging from 15% to 25%, which ensures valid information upon decoding. It enables the encoded data to be 100% restored even if up to 35% of the symbol becomes damaged or is missing.

The EDAC feature not only determines whether an error is present in the code but also assures that the correct information is recovered during the decoding process. Accordingly, a VeriCode symbol maintains its integrity by delivering accurate information or no information. The error correction is implemented by Reed–Solomon encoding.

Omnidirectional reading and skew resistance

VeriCode can be read in any orientation, and no manual positioning of the code is necessary. The omnidirectional feature allows the symbol to be used for applications where symbol orientations are random, as with the parts in an automated assembly process. The skew versatility of VeriCode allows the code to be captured and decoded at an up to 60% angle away from the horizontal plane. This feature, along with its error correction capability, enhances VeriCode's reading robustness.

3.3.5   Advantageous features

Marking capability

The VeriCode symbol can be encoded and marked on approximately 98% of all materials, since over 35 different marking methods are available. Some examples of such marking methods are presented in Table 3.4.

Integration with other automatic identification technology

VeriCode can be used with or within other electronic media such as RFIDs or smart chips for data storage and access. Data within an RFID or smart chip can be modified without being recognised. A VeriCode symbol that contains a user's personal or secret

Table 3.4. VeriCode marking methods

Marking method	Description
Thermal printers	Produces labels on heat sensitive paper
Laser etching	The most versatile VeriCode marking method
Dot peen	Marks by creating small indentations directly on the material surface
Micro-abrasive blasting	A non-contact process that uses abrasive particles flowing in a fine stream of air to engrave the VeriCode symbol on the material's surface

information can be embedded into these technologies (e.g. as an RFID tag) and can add another level of security.

Biometric data

VSCode is distinctive in being able to store large amounts of biometric data. This can be secured with techniques such as unique data encryption plus custom design capability. Once the encrypted data are incorporated within the VSCode symbol and printed on an ID card they cannot be changed, unlike in other technologies (e.g. RFID). The VSCode symbol can be inspected visually for signs of tampering or damage.

3.3.6 Scanning and decoding

VeriCode symbols can be scanned and decoded using CCD camera technology.

3.3.7 Applications

With VeriCode and VSCode, the mobile phone serves the function of an electronic credit card or ID card. Using the electronic media, information such as credit card details, bank account, airline tickets, bus tickets and medical information can be transmitted to the individual mobile phone. For example, if a concert ticket is purchased, it can be sent to the mobile phone of the customer in the form of VeriCode and can then be used in the same way as an ordinary ticket is used: the customer is required to show the symbol at the entrance to the concert hall. Once the code is decoded and the validity of the ticket is checked, the customer is allowed to enter the hall. This new type of mobile phone application protects privacy, prevents fraud and provides a secure and quick way to handle daily transactions.

3.4 mCode

3.4.1 Overview

The mCode[7] symbol was specifically developed to meet the needs of emerging camera phone applications and specifically to maximise the data capacity within a given space.

---

[7] The material on mCode was referenced from [32] (US Patent 7 412 089), unless otherwise indicated.

It has a unique finder pattern and special compression scheme for encoding URLs. The finder pattern is made up of a combination of unobtrusive dot-like cells (called blobs) and requires less space than other 2D barcodes. More importantly, mCode can be integrated with eye-catching commercial or artistic images such as a company logo.

The mCode system was designed for a diverse range of users with different interests. For example, mCode symbols can be used as public codes that anyone in the world can access. Depending on user requirements, however, mCode can be customised for a predefined group of users, providing them with a customised code reader.

The mCode's high data capacity allows it to be used as a database 2D barcode. Furthermore, the mCode reader is designed to decode the codes on the phone itself, requiring no access to a backend server. However, the mCode system can support customised business logic that provides connections to backend servers to resolve code content or retrieve data when requested.

Using a publicly available Web-based mCode generator, users can create their own mCode symbols, encoding URLs, SMS messages, contact details or phone numbers.

3.4.2   mCode symbol structure

The mCode's unique finder pattern is constructed from a flexible configuration of elements with recognisable attributes. The cells that compose an mCode symbol are called blobs. Any suitable shape or design such as a circle, rectangle, square, ellipse, pentagon, star, icon or branded shape may be used as a blob.

A standard mCode symbol has triads of blobs at three corners and a single 'solo' blob at the remaining corner (see Figure 3.19). A triad is made up of three blobs, a root and two equivalent arms, which are at a predetermined location in the symbol. The triad and solo locations can form corner positions of various shapes: L-shape, triangle, regular polygon,

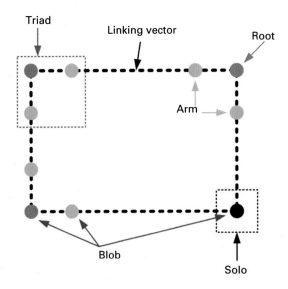

Fig. 3.19   Symbol structure of mCode.

irregular polygon or oval. Figure 3.19 shows a rectangular mCode finder pattern, in which adjacent triads and the solo are connected by implied 'linking vectors'.

A linking vector is formed by the observed collinearity of any two blobs in a triad with another blob, either a solo or a blob from a different triad. Three randomly occurring blobs in an image are rarely collinear. The collinearity of three points is easily and efficiently detected by a simple computation with location coordinates. Collinearity is also a property preserved under perspective distortion. Thus, a linking vector relationship is required between some or all of the triads and solos comprising a finder pattern, and this creates an efficient selection criterion in a finder-pattern search. Also, a linking vector could involve the required collinearity of more than three blobs and/or could require the collinearity of three blobs which are each from different solos or triads. The implied linking vectors illustrated in Figure 3.19 by broken lines are constructed from three triads and one solo, forming a rectangle.

3.4.3 Symbol description

Symbol size, shape and colour

The mCode's greatest advantage is its flexibility. There are currently more than a thousand different sizes of mCode available, any of which can carry exactly the same information up to the capacity of the particular size. One decoder algorithm detects any and all sizes simultaneously, without a noticeable overhead for the number of different sizes. The maximum symbol size of currently used mCode is 44×44 modules.

In addition to the standard square or rectangular symbol shapes, mCode has been produced in other shapes, such as an L-shape, which are useful for branding. Fundamentally, the patented finder pattern for mCode allows the design of codes with arbitrary shapes, data-module or aesthetic-module intermixing and full-colour options. The basic mCode is monochromatic. However, colour elements can be used as one type of finder pattern. In short, mCode can be any polygonal shape in any colour with flexibility in size.

Data capacity and encodable character set

The finder pattern for mCode occupies a lower percentage of the total symbol size than any other standardised or proprietary 2D barcode design, resulting in an optimised data capacity within the given space: mCode has substantial density, especially for small-capacity codes.

The core of mCode is a binary data carrier that can carry any number of bits up to a maximum for the particular symbol size. As a result, it can support various character sets, ranging from alphanumeric character, binary and Unicode to any other formats depending on user requirements. The maximum mCode symbol can carry approximately 150 bytes of data.

Error correction

Reed–Solomon technology is used to increase the robustness of mCode. The error correction capability of mCode is comparable with Data Matrix or mid-range QR Code error correction. Furthermore, an independent error detection technique is used in mCode encoding. It efficiently prevents a false decoding.

Most 2D barcodes use a 'one size fits all' approach for their error correction, resulting in wasted bits after error correction blocking, in addition to wasted bits on average in the base message after compression. Unlike other 2D barcodes, mCode uses a variable-size error correction polynomial, which depends on the exact size of the code. This enables mCode to waste fewer bits, which results in greater code capacity. One numeric digit can be compressed into fewer than four bits.

3.4.4   Advantageous features

Design flexibility

Graphics can help brand a code or can indicate its intended purpose. One particularly useful application is to add graphical elements in and/or around a code that can serve several purposes. Being flexible in position and taking up much less space than other barcodes, the unobtrusive finder pattern of mCode makes it easy to integrate mCode symbols with any advertisement graphics or consumer-friendly branding images, while maintaining the integrity of its overall appearance.

The data elements, and possibly the finder pattern elements, might be positioned so as to form parts of human-recognised symbols such as numbers, letters, words, symbols or icons. Figure 3.20 presents an example of mCode in which a graphic is embedded.

Data compression

mCode can implement compression for the common data formats. The standard text compression of mCode is similar to that used in QR Code and Data Matrix code, with the same compression ratios. In addition to the common compression technique, mCode implements a special form of compression for URLs, which results in compression ratios

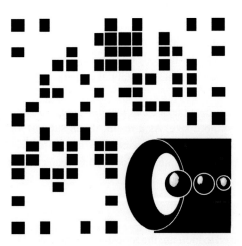

Fig. 3.20   Example of an mCode symbol with a graphic (adapted from [32]).

superior to those of other 2D barcodes. mCode also uses a superior form of compression for all numeric content.

3.4.5 Scanning and decoding

Once an mCode symbol is captured by a camera phone, the following main steps will occur to retrieve the original data.

(i) The image is processed; this includes pre-processing.
(ii) The blobs in the image are located and extraneous blobs are removed.
(iii) Potential triads are located.
(iv) Triad and solo candidates in the image are matched to the triads and solo in the specification.
(v) The collinearity of the linking vectors is checked.
(vi) The homography is calculated.
(vii) Decoding and other types of processing takes place.
(viii) Thus the raw data are retrieved.

3.4.6 Applications

The mCode system enables a wide range of applications based on code scanning and the resultant Web link establishment, ranging from private services such as direct URL encoding, the short message service (SMS), email and contact information to commercial applications such as the integration of the mCode system with systems for mobile commerce and mobile payments.

3.5 Trillcode

3.5.1 Overview

Trillcode,[8] which was developed by Lark Computer, a Romanian company, is a database 2D barcode designed for use with mobile phones. One of the advantages of Trillcode is its scalable design, which not only increases data capacity but also allows internal modifications. As in mCode, an attractive image and/or text can be integrated easily into a Trillcode symbol, to make it an eye-catching advertisement. The high data capacity allows Trillcode to encode a sound file that can be decoded by the reading software installed in a mobile phone, enabling one to play a melody on the phone.

3.5.2 Trillcode symbol structure

Each Trillcode symbol consists of two essential components, a finder pattern and a data area. The finder pattern, which lies on the perimeter of the symbol, is made up of a

---

[8] The material on Trillcode was referenced from [33], unless otherwise indicated.

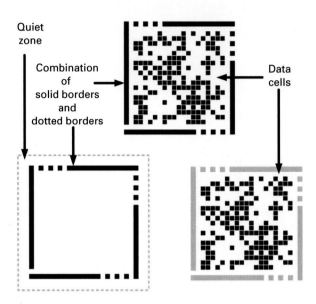

Fig. 3.21 Trillcode symbol structure.

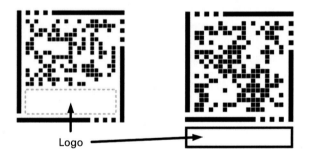

Fig. 3.22 Trillcode symbols with logos, inside the symbol (left) or outside the symbol (right).

solid border and three partly solid and partly dotted borders (see Figure 3.21). The data area comprises dark square data cells and spaces in a lighter colour. There are thin gaps between the dark cells, separating them slightly.

Optionally, a Trillcode symbol can include a data-free area either inside or outside the symbol. This space is suitable for the placing of graphic images, such as a company logo either in image or text format (see Figure 3.22). The entire symbol is surrounded on all four sides by a quiet zone.

3.5.3 Symbol description

Symbol shape and size
The basic shape of the Trillcode symbol is a square. However, it can become a rectangle when an image and/or text is placed outside the symbol. Trillcode, which is a database 2D

Fig. 3.23   Examples of defective Trillcode symbols.

barcode, is scalable in size. Between eight and 210 numeric, alphanumeric and special characters can be encoded in a Trillcode symbol.

Symbol colour
Any combination of dark and light colours can be used to encode data. However, in general, black and white are used to create a Trillcode symbol.

Error correction
With the use of error correction codes, Trillcode symbols that have defects such as scratches can be read successfully. Figure 3.23 presents examples of defective Trillcode symbols that are nevertheless legible.

3.5.4   Advantageous features

As with mCode, the design capability of Trillcode allows users to brand their symbols by inserting an image or text. This not only indicates the owner and the intended purpose of the symbol but also improves its visual attractiveness.

3.5.5   Applications

Trillcode offers users various applications ranging from displaying an advertisement, navigating a Web site, sending an SMS, dialling a phone number, sending an email, sending forms and adding contacts or events to the phone lists to playing music [33]. A Trillcode symbol with an animated image is eye-catching and can be a strong marketing tool. The great data capacity of Trillcode allows users to use it as an entertainment tool. A melody can be encoded into a Trillcode symbol, which can be decoded by a camera phone with reading software and be played on the phone.

One unique application is an order or feedback form embedded in the Trillcode symbol. It allows users to send order forms or obtain feedback via SMS or the wireless application protocol (WAP) using an input form embedded in the symbol.

Using the publicly available Trillcode encoder, not only companies but also an individual user can create their own personalised Trillcode symbol and use it according to their requirements.

3.6   ShotCode

3.6.1   Overview

The distinguishing feature of ShotCode is its aesthetic round shape. The symbol, originally called SpotCode, was developed by High Energy Magic Inc. to enhance human–computer interactions. SpotCode allows a Bluetooth-enabled camera phone to control active displays by acting as a sophisticated pointing device regardless of the shape and size of computer displays. The code can also work as a user interface for devices without a display or input capability of their own [34]. SpotCode recognises the rotation of symbols in the image and thus this barcode system is angle-free. In 2005, SpotCode was acquired by OP3 (OP3 AB Sweden, 2005) and named ShotCode. The function of ShotCode has been redefined as an index tag to link the physical and the digital worlds, thereby accessing remote databases.

3.6.2   ShotCode symbol structure

SpotCode[9] was a derivative of another circular 2D barcode or 'ringcode' (also known as a TRIP tag or TRIP code); TRIP stands for Target Recognition using Image Processing [20]. The TRIP code has been used with inexpensive CCD cameras (e.g. webcams) to track a moving target such as a person; such real-time location information can be used to provide a ubiquitous computing environment [20].

In order to describe the symbol structure, the original 2D barcode, namely, the structure of TRIP code, is introduced in this section. The symbol structure is provided in Figure 3.24.

In TRIP codes, a ternary number in the range 1 to $19\,683$ (i.e. $1 - 3^9$) is encoded using two concentric black and white rings surrounding a 'bull's eye'. As illustrated in Figure 3.24, these two concentric rings are divided into 16 sectors. The first sector (i.e. the synchronisation sector) indicates where the TRIP code begins. This sector presents alternating black and white ring sections to be used as a reference. Moving anticlockwise, the subsequent two sectors are reserved for storing an even-parity check on the encoded identifier (i.e. the TRIP code). The following four sectors encode the radius of the central bull's eye in millimetres. Finally, the remaining nine sectors encode a ternary identifier. For example, the value of the tag in Figure 3.24 is 102011221210001. The even-parity check is used to detect possible decoding errors.

3.6.3   Symbol description

Symbol shape
The reason why TRIP code is based on a circular pattern is that it was originally designed for the 3D localisation of tagged objects in highly cluttered environments. In such an environment, the central bull's eye of a TRIP code presents a very distinctive pattern: circles are less common shapes than right angles, squares and rectangles [20]. Furthermore,

---

[9] The material on SpotCode and TRIP code was referenced from [34] and [20], respectively, unless otherwise indicated.

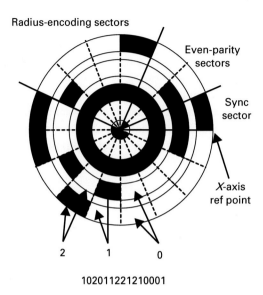

Radius-encoding sectors

Even-parity sectors

Sync sector

X-axis ref point

2   1   0

102011221210001

Fig. 3.24   The TRIP code symbol structure.

the detection of squares in such cluttered environments is an expensive computational task given the many straight-edge combinations possible within an image.

Symbol colour

Like in most currently available 2D barcodes, a black and white pattern was used for TRIP code in order to provide a distinctive contrast between the marker and the surrounding objects. The use of black and white resulted from three major factors:

(i) processing monochromatic images is computationally cheaper than processing colour images;

(ii) currently, monochromatic printers are more widely available; and

(iii) using colour is both difficult and unreliable as the ambient lighting conditions considerably affect the reading of objects and, furthermore, the colour sensitivity of different CCD devices is not identical [20].

Error detection

The even-parity check that is allocated to the second and third sectors in a TRIP symbol is used to detect data decoding errors [35].

3.6.4   Advantageous feature: 360° orientation

The first sector (i.e. the synchronisation sector) in a TRIP symbol is used to specify its orientation. Hence, TRIP code can be read omnidirectionally [20].

3.6.5   Scanning and decoding

Whereas inexpensive CCD cameras such as webcams or closed circuit television (CCTV) cameras are utilised in a TRIP system, ShotCode uses the built-in CCD cameras of mobile phones as image capturing and processing devices. There may be a difference in quality between the devices used in the former system and in the latter applications. To detect and process a ShotCode symbol, tag-reading software should be installed in the mobile phone. The software decodes the symbol as soon as it appears in the viewer frame of the camera phone. The code is read in anticlockwise fashion, starting from the synchronisation sector.

3.6.6   Applications

Applications of SpotCode focus on the human–computer interaction, whereas ShotCode is often used for *electronic commerce* (e-commerce). The following are examples of applications where SpotCode and ShotCode are utilised, respectively.

SpotCode application

An example of SpotCode applications is World Map. World Map is designed for travel agencies to allow users to access flight information via a mobile phone [34]. In this application, a user interacts with a plasma screen that shows a map augmented with SpotCodes in order to book a flight ticket. This is implemented in the same way as a computer and a mouse are used. Since a camera phone also has a display screen, the relevant information is shown on either the plasma screen or the camera phone display.

ShotCode applications

ShotCode allows a user to interact with Web contents using a Web link established via ShotCode and a camera phone. These applications include business cards, simple games, on-line shopping, trade-fair information, time registration, logistics and individual communications [36].

3.7   Visual Code

3.7.1   Overview

Visual Code[10] was originally developed to enable human–computer interactions using camera phones. The novel features of Visual Code are its code coordinate system, its visual detection of phone movement and its determination of the mobile-phone-camera rotation angle and amount of barcode tilt. The code coordinate system is used to map arbitrary points in the image plane to corresponding points in the code plane. The movement-detection algorithm allows the mobile phone to function as an optical mouse.

---

[10] The material on Visual Code was referenced from [17] unless otherwise indicated.

Owing to the mobility of camera phones, the scanned symbols can be randomly rotated and/or tilted. Visual Code can use the amount of rotation and tilting of a barcode in the captured image as additional input parameters. Hence, with the integration of these algorithms the Visual Code system allows users to select the target item in the display image precisely, to obtain and find different types of information by rotating the phone and to update it continuously as the phone is moving. These features enable applications that support real-time interactions between the camera phone and nearby active displays.

Although its data capacity is rather limited, Visual Code can function both as a portable database and as an index to a remote database.

3.7.2 Visual Code symbol structure

Figure 3.25 shows the Visual Code symbol structure and its code coordinate system. It can be seen that a Visual Code symbol is composed of larger and smaller guide bars, three corner stones and a data area with the actual code bits [17]. The guide bars are used for determining the location and orientation of the barcode and also for detecting distortion. They are able to detect even strongly tilted codes. The data capacity in the data area is 83 bits [17]. The three corner stones are designed for the code coordinate system. Each barcode defines its own local coordinate system, with origin at the upper left corner of the barcode and one unit corresponding to a single code-bit element [17].

Depending on the barcode size, the mapping between points in the image plane and points in the barcode plane has a greater precision than one coordinate unit. The x-axis extends horizontally to both the left and right beyond the symbol itself. Correspondingly, the y-axis extends vertically beyond the top and bottom edges of the symbol. For each barcode found in a particular input image, the code-recognition algorithm establishes

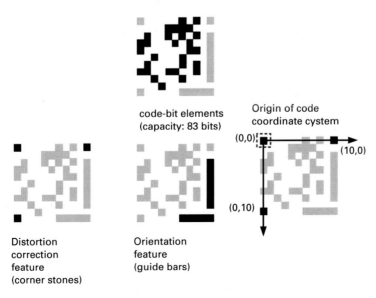

Fig. 3.25   Visual Code symbol structure and code coordinate system.

a bijective (i.e. one-to-one and onto) mapping between arbitrary points in the barcode plane and corresponding points in the image plane [17].

3.7.3 Symbol description

Visual Code is a square 2D barcode that consists of black and white square cells. The code can encode decimal or hexadecimal numbers. The code bits are protected by an (83, 76, 3) linear code that generates 83-bit codewords from a 76-bit value and has a Hamming distance of 3 for the detection of errors and false-orientation features [17].

3.7.4 Advantageous features

360° orientation

The location and orientation of a Visual Code symbol can be identified using the guide bars. Thus, the barcode can be read omnidirectionally. Since Visual Code uses the amount of rotation and tilting of a symbol in the image as additional input parameters, the symbol-tracking capability of Visual Code is strong and continuous [37].

Human–computer interaction capability

The code coordinate system, rotation, tilting and visual-movement detection are integrated into the Visual Code system. These features enable applications such as user interaction with nearby active displays [17]. The recognition algorithm can accurately map the coordinates of a targeted point in the code coordinate system. Such operation is independent of the orientation of the camera relative to the symbol (e.g. distance, rotation, and tilting) and also independent of the camera parameters (e.g. the focal point and focal distance). This allows each point in the viewed image to be associated with specific operations. That is, a single Visual Code symbol can be associated with multiple areas and, furthermore, a single image point can be associated with multiple information aspects provided by the various rotation and tilting angles.

3.7.5 Scanning and decoding

The decoding process of a Visual Code symbol involves two steps: Visual Code recognition and symbol reading. Code recognition can be achieved by the following recognition algorithm.

Recognition algorithm

Once the image of a Visual Code symbol is captured by the built-in camera of a mobile phone (i.e. as input), the following main steps are performed by the recognition algorithm of the Visual Code reader. The output is code information for each detected symbol. This includes:

  (i) the code value;
  (ii) the image pixel coordinates of the corner stones and guide bars;
  (iii) the rotation angle of the code in the image;

(iv) the amount of horizontal and vertical tilt;[11]

(v) the distance of the camera to the code;

(vi) a projective warper object (i.e. parameters to correct symbol distortion) for the code, which implements a planar homography used to transform image coordinates to code coordinates and vice versa;

(vii) the width and height of the originating image; and

(viii) a flag indicating the result of error checking [17].

Symbol reading

Once position detection of the guide bars and corner stones has been completed and a suitable projective mapping (also known as a projective or homogeneous transformation)[12] has been computed, the next step is reading the code. The encoded data can be retrieved by interpreting the black and white pixels in the data area of the Visual Code symbol. The resultant data can be used as additional input for a human–computer interaction application.

3.7.6 Applications

As for SpotCode mobile applications, Visual Code is utilised in applications where human–computer interaction is performed. An example is a weather forecast electronic-newspaper page that contains Visual Codes (see Figure 3.26).

The 17 regions on the map and the entries in the table are individually mapped to different URLs and thus hyperlinked to specific online content. This enables a user to

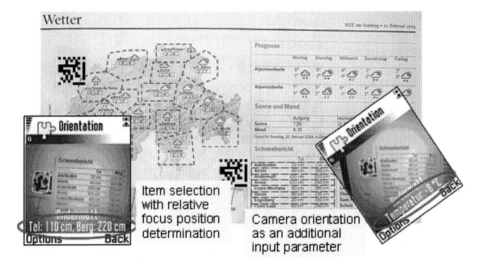

Fig. 3.26 Example of Visual Code weather forecast application (adapted from [17]).

---

[11] The term 'tilt' denotes the inclination of the image plane relative to the code plane. The horizontal tilt is the inclination of the image plane relative to the horizontal axis of the code. Likewise, the vertical tilt denotes the inclination of the image plane relative to the vertical axis of the code [17].

[12] This will be explained in detail in Section 6.4.1.

obtain real-time information about the forecast by accessing the relevant Web site through the Visual Code selected. Furthermore, users can see different aspects of the online information by rotating their camera phone. In Figure 3.26, a vertical orientation provides information about the snow depth at the chosen area, while the current temperature can be seen by slightly rotating the camera phone.

Other applications provided by the Visual Code system include a camera-controlled wireframe model of a house, a Pong game that can be controlled by tilting the wrist to the left or right and large subway maps that are scrolled in response to phone movements. In such applications, a camera phone is being used as an optical mouse.

3.8   BeeTagg

3.8.1   Overview

BeeTagg[13] is an index 2D barcode invented by Maass and Meyer in Switzerland in 2007. It is optimised to be processed by mobile devices such as mobile phones and personal digital assistants (PDAs). That is, it is designed to provide a fast and robust reading of barcode symbols using devices with limited capabilities for use as barcode scanners.

3.8.2   BeeTagg symbol structure

A BeeTagg symbol consists of three major components, a closed solid border, a data area (known as a pixel pattern) and a data-free area (also called a pixel-free area). Figure 3.27 presents the BeeTagg symbol structure.

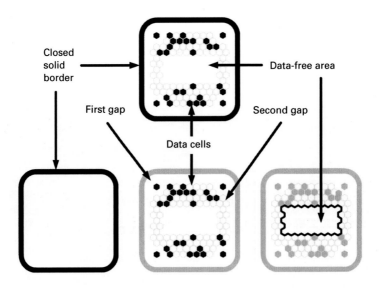

Fig. 3.27   BeeTagg symbol structure.

---

[13] The material on BeeTagg was referenced from [38] (PCT/EP2007/052544), unless otherwise indicated.

The closed solid border or outer frame works as a finder pattern and is used to locate the BeeTagg symbols. The data area comprises hexagonal data cells arranged in a honeycomb-like manner. The data cells often include format information such as the shape and location of the data-free area in addition to the encoded data.

There is a gap between the closed solid border and the data area, called the first gap. The data cells are also separated by a small gap of constant thickness that avoids their direct contact; this is called the second gap. The width of the second gap is 5%–10% of the smallest diameter of a data cell. These gaps facilitate the fast and robust detection of BeeTagg symbols. The data-free area located at the centre of each symbol is suitable for the placing of an image or text (e.g. a company logo) to provide additional information about a symbol and/or improve its visual attractiveness.

3.8.3 Symbol description

Symbol shape
The shape of a BeeTagg symbol can be rectangular, round or hexagonal; however, a square is most commonly used. Like the shape of the symbol, the shape of the data cells can differ depending on the application. However, a hexagonal cell shape is commonly used since it allows fast and robust symbol detection. Hexagonal cells of a uniform dark colour (e.g. black) represent a logical 1 whereas a thin outer line surrounding an area of lighter colour (e.g. white) represents 0.

3.8.4 Advantageous features

Attractive symbol design
Each BeeTagg symbol offers free space where additional information can be integrated in either a form of image or text such as a company logo. This allows a BeeTagg symbol to be attractive and identifiable. BeeTagg can also be personalised: depending on its layout, it is possible to arrange the data-free area in different shapes and in different locations (i.e. either the centre or along at least one edge of the closed solid border). Figure 3.28 is an example of a personalised BeeTagg symbol.

Robust symbol-location capability
The thick closed solid border of a BeeTagg symbol, which works as a finder pattern, is easily distinguishable. It allows accurate symbol location even in a challenging environment where a BeeTagg symbol is surrounded by printed letters or pictures (e.g. a BeeTagg symbol placed in a magazine).

3.8.5 Scanning and decoding

The BeeTagg code can be read from a single image or from one or more frames of a video stream from a camera.

The BeeTagg decoding procedure includes the following steps:

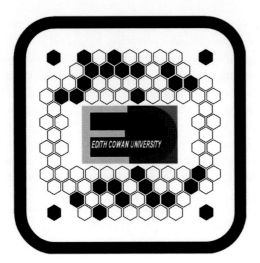

Fig. 3.28   Example of a personalised BeeTagg symbol.

(i) conversion of the scanned image into a greyscale image, which is then thresholded to a binarised image (i.e. a black and white image);

(ii) identification of the closed solid border (i.e. the finder pattern) of the target BeeTagg symbol;

(iii) location of the positions of the data cells in the four corners using the secant method;

(iv) computation of the parameters of the transformation function using the corner data cells;

(v) application of the transformation function to the BeeTagg image to correct its distortion; and

(vi) decoding of the data cell values.

The correction of the symbol distortion is performed using the data cells in the four corners of the BeeTagg symbol (see Figure 3.29).

In the BeeTagg secant method, two points that are equidistant from the middle point of an imaginary line connecting them move along the closed solid border of the symbol. When the equidistant points approach a corner of the closed solid border, the imaginary line temporarily runs as a *secant* across the first gap between the closed solid border and the data area. The corner data cell is arranged in such a way that when the secant line reaches its innermost position it runs through a corner data cell (see Figure 3.29). This event can be identified by analysis of the intensity distribution along the secant. When the line is moving along a straight segment of the closed solid border, its middle point also lies on the closed solid border. Conversely, when the line forms a secant around a corner, its middle point crosses the first gap until it reaches and touches the corner data cell. Since the relative position of the closed solid border is known, the position of the corner data cell can be computed using this secant method.

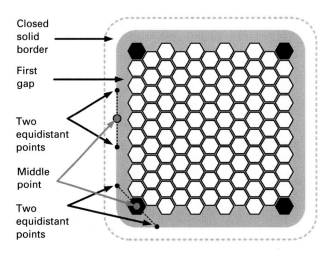

Closed solid border

First gap

Two equidistant points

Middle point

Two equidistant points

Fig. 3.29   Schema of the four-corners determination process.

The position of the centroid of the corner data cell is then determined using an averaging method. From the centroids of the data cells in the four corners, distortion correction of the scanned BeeTagg image is performed. By this operation, the data cells that are arranged within the four corners (see Figure 3.29) are also aligned correctly, which enables accurate reading of the BeeTagg barcode.

Once the distortion of the scanned symbol is corrected, the encoded data are retrieved. Since BeeTagg is an index 2D barcode the retrieved data are redirected to an application server, which then returns the information to the designated device (e.g. a camera phone) for further processing.

3.8.6   Applications

As for other index 2D barcodes, BeeTagg provides fast and easy access to mobile phone contents, allowing users to interact with the contents via BeeTagg and the camera phone. Owing to its flexible design, BeeTagg can also be used as an attractive and/or eye-catching advertisement medium. For example, Swiss Post uses BeeTagg to make contact with potential apprentices [39].

3.9   ColorCode

3.9.1   Overview

In 2000, Professor Han and his team at Yonsei University, Korea, developed Color-Code,[14] a colour 2D barcode designed for mobile phone applications. This is perhaps

---

[14] The material on ColorCode was referenced from [40] (US Patent 7 020 327), unless otherwise indicated.

the most eye-catching visual tag to date, while being flexible in design and form. The number of cells per symbol is set flexibly according to the desired data capacity. Hence, it is possible to use ColorCode as a portable database as well as an index to a remote database. Depending on the user's requirements and/or the shape of the medium on which the symbol is printed, the cell shape can be circular, oval or polygon; however, symbols constructed from a matrix of rectangles are preferred.

A colour image could be reproduced differently from its original colour, depending on the image capturing and processing device (e.g. a scanner, digital camera or webcam), the printing devices used and the quality of substrate on which the image is printed. Lighting conditions can also have a great impact on the colour value of a captured image. To overcome this problem, ColorCode uses reference cells which provide the standard colour and/or hue for correctly distinguishing each reproduced cell. The value of each colour and/or hue of cells in the data area is determined in relation to the value of the standard colour in the reference cells. Since the relative difference in colour (and/or hue) of a cell from that of the standard colour (and/or hue) is consistent regardless of the devices used for image capture or display, the data can always be correctly retrieved.

It is often believed that, in order to read colour symbols, expensive, high-resolution, cameras are required. However, ColorCode can be decoded by camera phones with relatively low camera resolutions such as VGA (video graphics array, i.e. 640 × 480 pixels) [37].

Originally, two different types of code, namely, GrayCode and Numeric Code were also available. These codes worked just as ColorCode does. A 10-digit numeric code was an option for mobile phone users without cameras. These are no longer in use.

ColorCode uses the colour element in addition to its two spatial dimensions for encoding data. Therefore, ColorZip™, the proprietor of ColorCode, classifies the code as a three-dimensional (3D) barcode.

3.9.2   ColorCode symbol structure

A current standard ColorCode symbol comprises a matrix of 5 × 5 cells rendered in a combination of four different colours, namely, black, blue, green and red.

A ColorCode symbol consists of at least two areas, namely the data area and a parity area. In addition to these two areas, a reference area and a control area can be included in a ColorCode symbol. Either a row-parity or a column-parity check or both row- and column-parity checks can also be included. With a position cell, a triple-parity check can be performed. The symbol structure of ColorCode is presented in Figure 3.30.

A reference cell provides a reference colour or reference hue for judging the colour or hue of a data cell formed in the data area. A control cell indicates a command or service for controlling target information about the data area such as the location of the reference cells, the location or properties of the parity area and the sequence for decoding the cells in the data area. The reference area and control area can be integrated in the area that is not used for data and parity cells.

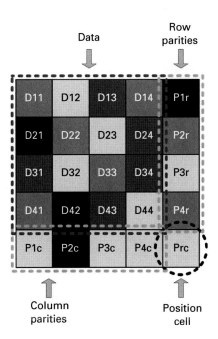

Fig. 3.30   ColorCode symbol structure (adapted from [41]). The row-parity check cells are P1r–P4r and the column-parity check cells are P1c–P4c.

3.9.3   Symbol description

Symbol shape

A standard ColorCode symbol is a matrix of 5 × 5 cells rendered in the combination of four different colours. Initially, 5 rows × 8 columns ColorCode in the same colour range was offered as an alternative. Currently, however, only the standard version is in use. Unlike the symbol shape, cells can be in a variety of shapes such as circular, oval or polygon.

Symbol colour

A combination of four different colours, black, blue, green and red, is normally used to generate a ColorCode. However, a different set of colours can be used.

Encodable character

Despite its potential for higher data capacity, ColorCode has only been used as an index code, encoding 10 digits.

Error detection

As presented in Figure 3.30, ColorCode includes parity areas that contain error parity checks to detect any incorrect colour recognition, which the system can then correct. The result of an exclusive OR (XOR) of code values in each column or row becomes the code value of the parity cell for the respective column or row and is converted to its corresponding colour value in the parity area of the symbol.

3.9.4   Advantageous features

Needless to say, the most advantageous feature of ColorCode is its eye-catching and flexible design. Examples of ColorCode symbols are presented in Figure 3.31. ColorCode decoding only requires 40% visibility of an individual cell. Thus, a barcode can be easily incorporated into company logos or other graphic designs, making the most of the remaining 60% of the space on the actual symbol.

Recently, in [41], a new type of such 2D barcodes, namely, a mixed code, was introduced (see Figure 3.32). A mixed code can be produced by placing a 2D barcode on top of another barcode. Not only 2D barcodes (e.g. QR Code, Data Matrix or PDF417) but also letters, logos or 1D barcodes can be integrated into a mixed ColorCode. Being

Fig. 3.31   ColorCode design variations (adapted from [41]). For a colour version, please see Plate 1.

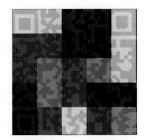

Fig. 3.32   Examples of mixed ColorCode (adapted from [41]). For a colour version, please see Plate 2.

visually appealing and with the facility to provide additional information, mixed codes can be an effective means for advertisement.

3.9.5 Scanning and decoding

ColorCode can be read from a distance of more than 5 meters by camera-equipped mobile phones, PDAs, PCs and bouses.[15] It works with a CCD or complementary metal oxide semiconductor (CMOS) camera having only 100 000-pixel resolution. It can even operate under shade, low-light or low-contrast conditions if at least 40% of every cell is visible.

3.9.6 Applications

Owing to its eye-catching design, ColorCode itself can work as a portable advertisement, providing camera phone users with a link to a designated Web page at the same time. This allows ColorCode to be a strong marketing tool. As a result, ColorCode is often used in e-commerce activities such as a new product-launch campaign.

3.10 High Capacity Color Barcode

3.10.1 Overview

Using only two colours whose colour distance is maximal, most existing monochromatic barcodes ensure robust reading capability even with readers that are of relatively low cost or resolution. However, such black and white barcodes are challenged by the need to increase data capacity within a given space.

Applying colour elements in encoding data, ColorCode could increase data capacity but, typically, its colour element has been used in index 2D barcodes to produce eye-catching symbols rather than to increase data capacity.

High Capacity Color Barcode (HCCB)[16] was developed to increase data capacity, thus overcoming the limitations of encoding data in the square or block matrix format using monochromatic colours. It can be generated in greyscale or colour tones at 8, 24, 32, 48 or other bit depths, depending on the application. By encoding a triangular high-density geometric symbol set, HCCB can achieve at least three times the density of industrial-standard 2D barcodes such as PDF417 or Data Matrix. With this improved data density, HCCB symbols can be used for transaction or identification applications, which demand a greater overall information content.

The built-in error detection and correction capabilities in HCCB enable the robust reading of small data cells or blurred symbols, which, in turn, allows HCCB to achieve at least 1100 bytes or 3300 symbols per square inch even when it is printed with a

---

[15] A bouse is a mouse with a built-in camera that scans ColorCode.

[16] The material on High Capacity Color Barcode was referenced from [42] (US Patent application, 20050285761), unless otherwise indicated.

conventional inkjet colour printer [42]. White separators are embedded between adjacent symbols to enhance the accuracy of symbol detection.

A similar approach to that taken by ColorCode is used in HCCB to overcome the difficulty in reading colour symbols. When compared with monochrome, colour is more susceptible to variations introduced by the particular printing device, the printing technology used (i.e. inkjet, colour laser or dye submission) and the printed medium, such as paper, and its ageing. The resulting colour value (i.e. output) may therefore deviate from the colour value that was sent in digital form to the printing device. Ageing of the output colour may also affect the printed colour values such as the colour tone, size or shape: the materials used to imprint symbols may fade, smear or absorb moisture over time and, as a result, change the colour value.

In High Capacity Color Barcode these difficulties are overcome by the incorporation of a reference palette within the symbol. It allows a comparison between the scanned, or sampled, symbol colours and a set of reference colours so that colour correction can be performed.

3.10.2   HCCB symbol structure

An HCCB symbol is made up of triangular cells, the white spacing between them and the reference palette. The symbol structure of HCCB is presented in Figure 3.33. Four different colours are used in this example symbol. However, the number of colours may differ depending on the user and/or application requirements, such as the space availability for a symbol and the amount of data to be encoded in the HCCB symbol.

3.10.3   Symbol description

Symbol shape and size
The HCCB's symbol shape is a triangle. Each symbol contains data to be encoded together with its cyclic redundancy check (CRC) value, additional Reed–Solomon codes and the

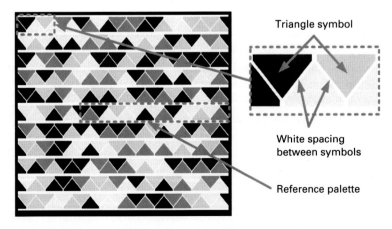

Fig. 3.33   Symbol structure of HCCB. For a colour version, please see Plate 3.

number of cells occupied by a reference palette. The symbol size is determined on the basis of the total amount of data to be encoded, including the CRC values, the Reed–Solomon code, the reference palette and the number of colours used, in other words, the number of bits that can be represented per encoded symbol.

HCCB data capacity

As mentioned above, by encoding a triangular high-density geometric symbol set that takes up less space than square cells, HCCB can achieve at least three times the density of industrial-standard 2D barcode such as PDF417 or Data Matrix within the given space.

Error correction

With HCCB, a CRC of the data to be stored is generated as an initial step in order to determine whether the encapsulated data have been successfully decoded in the scanning and verification phase [42]. In addition to the error detection scheme provided by CRC, HCCB uses Reed–Solomon error correcting codes to increase the robustness of HCCB symbols.

3.10.4 Advantageous features

Triangular cell format

The triangular cell format of HCCB is advantageous in terms of data capacity and accurate colour sampling. Needless to say, triangular cells occupy less physical space than square cells, resulting in an increase in data capacity. Furthermore, with only three straight-lined sides, triangles are less prone to the effects of anti-aliasing [42], which is often introduced during image capturing and processing procedures. The white space between cells further facilitates accurate colour sampling.

Multiple payloads

An HCCB symbol can store multiple individual items of data, each compressed as optimally as possible within the physical code, implementing essentially a small file system [42]. In order to achieve this by using existing barcode formats, special reader software that has knowledge about the data stored in the barcode is required.

Digital signing capability

The HCCB barcode has integrated digital signing features using state-of-the-art elliptic-curve-cryptography public key infrastructure (PKI) techniques [42]. The purpose of digital signing is to authenticate the integrity of the data stored in the barcode, ensuring that the original data are not compromised.

3.10.5 Scanning and decoding

The HCCB symbols can be scanned using a high-fidelity computer image-capture device such as a flatbed scanner, business card scanner, CCD based digital camera, video camera, Web camera or other close-contact scanning device [42].

Most existing barcodes use finder patterns such as identifiable guide bars and spaces or shapes that have distinctive features. However, an HCCB symbol is guide agnostic [42]. Once an HCCB symbol has been located within the captured image, operations to obtain the original symbol size and unrotated symbol position will be directly executed, given the four corners of the extracted symbol image, by using either trigonometric or vector-based arithmetic techniques [42].

When the correction of the captured symbol is finalised, an HCCB reader performs an absolute-position single-pixel sampling [42]. Then the sampled colour values of each cell are mapped to the colour values of each reference colour in the reference palette. If the reference palette has been damaged, the decoding software can fill in the gaps using a history of previously scanned palette colour values [42].

As the colour values of each cell are read, the bit pattern value for each cell colour sequentially regenerates the data byte encoded in the symbol. After error detection and correction have been performed by CRC operation and Reed–Solomon decoding, the remaining data should be the original data encoded in the HCCB symbol.

3.10.6   Applications

Biometric IDs, medical insurance or information cards often require large amounts of data capacity to encode personal information such as an iris scan, a fingerprint, a signature, a medical history, a DNA profile or other personal information [42]. Rather than resorting to expensive solutions such as a smart card or RFID, it is desirable to use inexpensive barcode symbols on a compact plastic, paper or another inexpensive medium. Making the most of the increased data capacity within a given space, HCCB symbols offer applications that can be used as security measures. For example, driving licences, passports or other ID media may require a high information content such as colour-digital face photographs. There is sufficient data capacity in HCCB barcodes for their use in such applications.

In April 2007, Microsoft Co. and the International Standard Audiovisual Number International Agency (ISAN–IA) announced an agreement whereby ISAN–IA has licensed HCCB technology for assistance in the identification of commercial audiovisual works such as motion pictures, video games, broadcasts, digital video recordings and other media.

3.11   Summary

Since the idea was first introduced, numerous 2D barcodes have been invented. These barcodes were initially developed for industrial use. However, the trend has changed as camera phones have infiltrated many aspects of our daily life. With the integration of CCD cameras, mobile phones have become networked image-processing devices. When used together with such camera phones, 2D barcodes work as a tag to connect the physical and digital worlds.

The potential capabilities of a combination of 2D barcode and camera phone technologies have attracted the attention of researchers and developers, and this has resulted in the development of the original 2D barcodes as well as novel mobile applications. Existing 2D barcodes such as QR Code and Data Matrix have been used to implement novel mobile applications. In fact, these two have been the pioneering 2D barcodes for implementing mobile applications, thus demonstrating their tremendous capabilities. Companies have also followed the trend and launched projects to develop systems that provide users with convenient and flexible services based on a Web link established via 2D barcodes and camera phones. Consequently, a variety of 2D barcodes exist.

The 2D barcodes used for mobile applications can be classified into two categories, database 2D barcodes and index 2D barcodes. The former refers to any 2D barcode that is scalable in data capacity and can operate as a robust and portable data file, whereas the latter works only as an index to a backend database. Database 2D barcodes include QR Code, Data Matrix, VeriCode, mCode, Trillcode and HCCB, while ShotCode, Visual Code, BeeTagg and ColorCode are categorised as index 2D barcodes.

Each 2D barcode has a unique symbol structure, encoding scheme, advantages, decoding algorithm and applications. A detailed description of each 2D barcode has been provided in this chapter. The development of 2D barcode symbologies is closely related to the evolution of barcode applications. This will be discussed in greater detail in Chapter 4.

4 Evolution of barcode applications

4.1 The improvement in capability of barcode technology

Two decades have passed since the first two-dimensional (2D) barcode was introduced in 1988. Over this period, barcode technology, especially that for 2D barcodes, has evolved continuously from its original tasks for automatic identification and data capture to a tool for realising augmented reality (AR) and/or ubiquitous computing. The focus of early researchers was on the capability of 2D barcodes as a whole, in comparison with the traditional one-dimensional (1D) barcodes, and the effect of their emergence on certain sectors of industry.

4.1.1 Taxonomic research on 2D barcodes and a comparison with 1D barcodes

Noting the potential of this new technology, Barnes *et al.* [15], for example, conducted thorough taxonomic research on the 2D barcodes available at the time. They first introduced 2D barcodes in comparison with 1D barcodes and then analysed, compared and contrasted the existing 2D barcodes with each other in terms of their capability (especially data capacity and accuracy), distinctive features and applications. As a result, Barnes *et al.* concluded that 2D barcode technology is more advantageous than its 1D counterpart in terms of physical size, storage capability and data accuracy, despite the cost of the increase in data density. These advantageous features allow a 2D barcode to be a robust portable database, while a 1D barcode can only work as a key to access a backend database for which network connectivity must be guaranteed.

Barnes *et al.* [15] also referred to the trends in certain industries towards standardisation of the use of 2D barcodes that have occurred at various times. The Automotive Industry Action Group (AIAG), for example, issued a policy for the industry that set the use of three specific 2D barcodes in three specific categories as a standard in 1995 [15]. The recommended 2D barcodes and relevant categories were Data Matrix in 'part marking and tracking', PDF417 for 'general applications' and MaxiCode for 'freight sorting and tracking'. Such attempts at standardisation in industry have facilitated the widespread use of certain barcodes. From a study of global markets and applications for 2D barcode reading equipment and systems, made by the research firm Venture Development Corporation, it was concluded in [15] that PDF417 and Data Matrix were more likely to dominate this market globally over a two- to three-year period, owing to the prevalence of the former and the rapid growth in adoption of the latter.

Two-dimensional barcodes were initially developed for applications where the use of 1D barcode was impossible owing to space limitations. Examples are unit-dose packaging in the healthcare industry and electronics assemblies in the electronics industry. The ability to encode a portable database, however, has made 2D barcodes attractive also in applications in which space is not at a premium [15]. Not only the industrial sector but also academic researchers in a variety of fields have had their eyes upon the potential of 2D barcode technology. Some of these have devised new applications, using existing 2D barcodes. Some have invented their own 2D barcodes to implement special tasks in their own mobile applications.

4.1.2 Predictable programs in barcodes

An interesting example of research that utilises an existing 2D barcode for a new type of application was [43]. The authors had conducted experiments on programming microwave ovens, using 2D barcodes as a means of delivering verified programs. This work demonstrated the feasibility of appliance programming through open application-programming interface (API) platforms. In these experiments, two issues of particular interest for open APIs on embedded devices were addressed: deliverability and predictability [43]. While the former concerns the means for delivering a program to an embedded device, the latter requires analytical techniques that make such programs predictable.

The Aztec code, a high-density matrix 2D barcode, was used to address the requirement of deliverability. Aztec code makes it possible to encode the program data in a limited space and deliver it to an embedded device. However, 2D barcodes are not suitable for the second purpose as they have no mechanism to analyse and correct data content, notwithstanding their capability to correct for physical errors such as spots or voids. Furthermore, it is difficult to check the content of a program once it is encoded in the form of a 2D barcode. In order to address this issue of predictability, a formal verification tool called Spin was adapted for this application. The main contribution of the research reported in [43] was that it demonstrated the feasibility of delivering verified programs in 2D barcodes, thus opening the door for open APIs to control a myriad of devices ranging from home appliances to medical devices. However, as was pointed out in [43], network connectivity was not considered in the proposed application, which limits the usability of such applications.

4.1.3 A novel 2D barcode: Secure 2D code

Secure 2D code, introduced in [44], is an example of a 2D barcode invented by academic researchers. Having discovered that there were no existing 2D barcodes that provided sufficient security, their focus was to increase the security and error correction ability of 2D barcode. Furthermore, these researchers proposed a pattern generation and reading system that pre-processes (e.g. compresses, encrypts, etc.) the data prior to encoding them. The applicable data range from different types of characters (e.g. Chinese characters as well as English) to graphic and audio.

To enhance security, a new data-hiding technique called 'data disguising' was introduced in [44]. In this technique the data are divided into two types, *general* and *secret*. The general data are first encoded in a 2D barcode format and then the secret data are embedded in the 2D barcode amongst the encoded general data. The resultant 2D barcode constitutes a Secure 2D code.

Unlike 1D barcode, originally 2D barcodes did not have any redundancy even though they are often exposed in an open environment where many types of physical damage (e.g. smudges, scratches and voids) may occur. In order to ensure the robustness and reliability of Secure 2D code in such environments, Reed–Solomon error correction is applied. The Reed–Solomon code also plays an important role in hiding the secret data within the encoded contents. In order for the system to provide a higher, layered, security, some error correction capability may need to be sacrificed [44].

These three topics, namely, security, error correction and pattern generation and reading, are addressed in [44]. As depicted in the name of the barcode, the emphasis in Secure 2D codes is all on the security and robustness of the barcode and its entire system. Hence, this code is especially useful in applications involving confidential data such as those in an identification card, driver's licence or passport. As in the case of the application devised by [43], no network accessibility is required by the Secure 2D code system.

4.1.4   Secure data transmission between the physical and digital worlds

INTACTA.CODE is another 2D barcode in which considerable emphasis has been put on security. It was originally developed by Intacta Laboratories in Israel for the defence industry, to enable the secure transmission of battle-field plans [45]. INTACTA.CODE offers secure electronic commerce (e-commerce) that includes paper to digital, digital to paper and digital to digital business transactions. The system has been developed on the basis that we are not living in a totally digital world. That is, many documents still start from paper sources and many end up in the same way [45]. Like Secure 2D code, INTACTA.CODE uses a variety of techniques (e.g. compression, encryption, error correction and encoding engines) for ensuring secure information transfer.

However, unlike Secure 2D code, INTACTA.CODE is designed on the notion that connectivity is always available worldwide in modern societies and that end-to-end security solutions across disparate systems are a must for data transmission processes, in order to ensure integrity of the format and content of the data. This barcode provides interoperability between paper and digital systems, on the assumption that almost seamless connectivity exists throughout the world regardless of the medium (whether physical or electronic). Hence INTACTA.CODE can traverse the border between the physical world and the digital world. However, INTACTA.CODE was not in fact developed to work as an interface or a visual tag to establish connectivity between the physical world and the digital world.

4.2   Two-dimensional barcode as a tool for ubiquitous computing

In contrast with the earlier researchers, researchers and inventors are now developing novel applications where barcode symbols are used as visual tags to establish network connectivity between the physical world and the digital world. These researchers often invent their own original 2D barcodes to implement specific tasks in such applications. These approaches can be roughly classified into two categories in terms of their final goals. In the first category are approaches in which the aim is to implement AR systems and in the second category are approaches in which the aim is to enable human interactions with the digital world.

More precisely, the aim of the latter approach is to use 2D barcodes as a means to achieve a world of ubiquitous computing where people use computers in their human environment without being aware of the computers' presence. It is a world where human-to-human interaction has a higher priority than human-to-computer interaction [46]. In such a world, human beings are not dependent on computers. Rather, computers exist to work for humans, making their life convenient, invisibly in the background [46].

Initially, new applications were designed in which barcode technology was integrated with wire-lined communication devices, such as PCs and inexpensive charge coupled device (CCD) Web cameras (webcams) and closed circuit television (CCTV) cameras. However, with the recent proliferation of camera phones, the focus of most new developments in 2D barcodes has shifted from wire-lined communication devices to wireless communication devices such as camera phones and personal digital assistants (PDAs).

The first 2D barcode invented for the AR environment was CyberCode. Other novel 2D barcodes such as TRIP code,[1] Visual Code and SpotCode (the successor of TRIP code) have been used as tools to make ubiquitous computing environments a reality. The newer 2D barcodes (i.e. Visual Code and SpotCode) were designed for wireless communication devices. However, the older CyberCode and TRIP code were developed for wire-lined communication devices because wireless devices were not common at that time.

4.2.1   Augmented reality

The Australian Communications and Media Authority (ACMA) defines the term *augmented reality* as follows:

[Augmented reality is] the opposite of virtual reality: instead of placing the user into a synthesised, purely informational, environment, the goal of AR is to augment the real world with information handling capabilities. An AR system generates a composite view for the user. It is a combination of the real scene viewed by the user and a virtual scene generated by the computer that augments the scene with additional information [47].

The notion of *virtual reality*, which attempts to build a world inside the computer, is considered to be diametrically opposed to the vision of a ubiquitous computing world

---

[1] Target Recognition using Image Processing code.

FieldMouse, type 1   FieldMouse, type 2

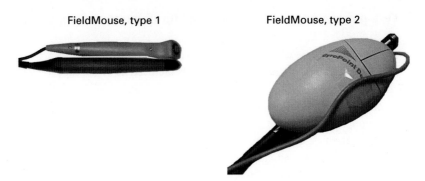

Fig. 4.1   Examples of FieldMouse technology. Type 1 is a combination of a pen-type mechanical mouse and a pen-shaped barcode scanner. Type 2 is a combination of a gyro-mouse and a pen-shaped barcode scanner.

[46]. Hence, AR can be seen as the polar opposite to the notion of ubiquitous computing, from a viewpoint that sees AR as an environment including virtual reality elements as well as real-world entities. However, to make such a ubiquitous computing world a reality, it is essential to implement a tagging identification (ID) system that establishes a link between the physical and the digital world, an issue which is often addressed in designing AR systems [7]. Hence, augmented reality can be regarded as leading to ubiquitous computing, and barcodes are used as one type of ID being tagged in the interface developed for AR systems.

FieldMouse with 1D barcode

The FieldMouse, whose invention was reported in [48], is an early example of a tagging ID system in which barcode technology was integrated with motion detection devices such as an optic mouse. As shown in Figure 4.1, a FieldMouse can take the form of either a pen-type mechanical mouse or a gyro-mouse integrated with a pen-shaped barcode scanner [48].

In this system, a traditional 1D barcode is used as an ID tag. Once an ID on paper or any other flat surface is detected by the scanner or barcode reader, its absolute location can be calculated by the movement of the mouse relative to the device's location of the ID. This enables a FieldMouse to work as a position-input device on any flat surface with a barcode printed on it. The FieldMouse system allows for a paper-based graphical user interface (GUI).

A point-and-click application called Active Book uses a FieldMouse as a position-input device. Embedded in each page of Active Book, are an ID tag encoded in a 1D barcode symbol and links to information on a computer. Once an arbitrary point on a page is detected by the FieldMouse, information corresponding to that point is retrieved from the computer via the embedded link. Although it is a paper-based book, it is interactive. By clicking arbitrary points users can, for example, hear the sounds or voice of characters in the book.

A paper-based GUI enabled by a FieldMouse can be used to control a device such as a television, as it can analyse the direction and amount of movement of the mouse.

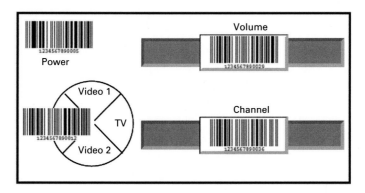

Fig. 4.2   A paper remote-controller with paper-GUI widgets (adapted from [48]).

For example, a user can control the volume of the television by sliding the FieldMouse over a 1D barcode symbol in which the digits that indicate different volume levels are encoded. Television channels can be controlled in the same way. Figure 4.2 presents an example of a paper-based GUI widget, a paper remote-controller.

With a portable digital display, a FieldMouse can create a two-dimensional AR system. While a FieldMouse, which works as a position-sensing device, keeps detecting arbitrary positions, the digital display is used to overlay computer-generated information about these positions to the real world. Such an AR system is used in the detection or tracking of electrical wiring behind a wall for maintenance. The AR system only works on a two-dimensional surface. Furthermore, the detected position is not always accurate [48]. Hence, possible applications of an AR system using a FieldMouse might be limited in comparison with more sophisticated three-dimensional AR systems.

However, applications based on the capability of the FieldMouse device (including two-dimensional AR systems) have demonstrated the usability and inexpensive operation of barcode technology in the field of AR and ubiquitous computing. As a consequence, the FieldMouse project has had a considerable impact on subsequent studies in these areas.

CyberCode

The workers in [7] also addressed tagging ID systems using a visual tag called Cyber-Code. CyberCode, introduced in 2000, was the earliest 2D barcode used in developing a visual tagging system for an AR environment. The major differences between CyberCode and the FieldMouse system are that a FieldMouse uses 1D barcode and can only be used on two-dimensional surfaces, whereas the newly invented two-dimensional CyberCode is capable of three-dimensional position tracking as well as two-dimensional position detection.

The shape of the CyberCode symbol is designed to compensate for any distortion of the captured image, so that the symbol can be recognised even when it is not perfectly placed in front of the camera. Figure 4.3 presents a CyberCode symbol and its structure.

The key features of the CyberCode are as follows.

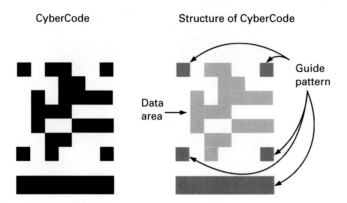

Fig. 4.3   The CyberCode tag structure.

(i) CyberCode can be recognised either by the low-resolution CCD or CMOS camera of a mobile device or by non-portable cameras such as a pan and tilt camera.
(ii) To maintain the robustness of the system regardless of the resolution of the image processing device used as a reader, CyberCode encodes fewer bits (i.e. 24 or 48 excluding error correction bits) than other 2D barcodes that existed at the time of its invention. Unlike the other 2D barcodes that were in use at the time, CyberCode was developed to work as an ID tag or index tag, assuming that it could access digital contents via 'network ready' mobile devices.
(iii) The CyberCode reader tracks the code position and orientation as well as the ID, so that the physical-world scene is correctly overlaid with the embedded augmented information. Switching of the overlaid information according to the IDs is straightforward since the data are encoded in the CyberCode symbol.
(iv) The 'low-resolution' approach of CyberCode allows CyberCode symbols to be displayed on computers and television screens, thus enabling human–computer interactions between the physical and the digital world. For example, a user with a CyberCode reader can retrieve the data encoded in a CyberCode symbol shown on the computer screen and use the data for its designated action.

The CyberCode recognition algorithm consists of two parts: recognition of the ID tag (i.e. the *visual tag recognition algorithm*) and determination of its three-dimensional position in relation to the camera (i.e. the *three-dimensional position reconstruction algorithm*).

Visual tag recognition algorithm
As illustrated in Figure 4.4, the CyberCode visual tag recognition algorithm operates in the following steps.

(i) The captured image is binarised using adaptive binarisation methods.
(ii) The connected regions that have a specific second-order moment are located; these are considered to be the candidate guide bars.

(i)   (ii)   (iii)   (iv), (v)

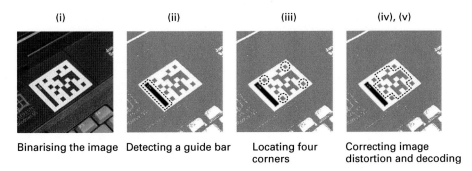

Binarising the image   Detecting a guide bar   Locating four   Correcting image
                                               corners   distortion and decoding

Fig. 4.4   CyberCode recognition steps.

(iii) The four corners of the marker region are located using the position and orientation of the guide bar found in step (ii).

(iv) The image distortion and tilt are corrected.

(v) The CyberCode tag is decoded after the algorithm has confirmed that the candidate image is a CyberCode tag by checking the error bits [7].

Three-dimensional position reconstruction algorithm

Once the designated CyberCode symbol has been recognised, the properties of the symbol are used to estimate the position and orientation of the camera. The four known points on the image plane, namely, the four corners of the CyberCode, enable the computation of a matrix representing the translation and rotation of the camera in a real-world coordinate system. Through this estimation, the algorithm minimises the following expression:

$$E = (\vec{v0} \cdot \vec{v1})^2 + (\vec{v1} \cdot \vec{v2})^2 + (\vec{v2} \cdot \vec{v3})^2 + (\vec{v3} \cdot \vec{v4})^2 + (\vec{v4} \cdot \vec{v5})^2, \qquad (4.1)$$

and this ensures that the estimated coordinate system is orthogonal [7]. In Equation 4.1, $\vec{v0}, \ldots, \vec{v3}$ are the orientation vectors of the four edges, and $\vec{v4}$, $\vec{v5}$ are the two diagonals of the code frame. The plane of the vectors $\vec{v0}, \ldots, \vec{v5}$ can be denoted by $\vec{n}$, a vector normal to the plane. Thus the minimisation constraint can be expressed as

$$E(\vec{n}) \rightarrow \min. \qquad (4.2)$$

The *downhill simplex method* is used to estimate the vector $\vec{n}$ that minimises E. Once $\vec{n}$ is computed, the vectors $\vec{v0}, \ldots, \vec{v5}$ can be recalculated in terms of $\vec{n}$. A point $(x, y, z)^T$ in the real world is related to the point $(X, Y, Z)^T$ in the camera coordinate system by a matrix:

$$\begin{pmatrix} X \\ Y \\ Z \\ 1 \end{pmatrix} = \begin{pmatrix} \vec{e_x} & \vec{e_y} & \vec{e_z} & \vec{e_t} \\ 0 & 0 & 0 & 0 \end{pmatrix} \begin{pmatrix} x \\ y \\ z \\ 1 \end{pmatrix}. \qquad (4.3)$$

Here $\vec{e_x}$, $\vec{e_y}$, $\vec{e_z}$ and $\vec{e_t}$ denote vectors of camera rotation coefficients and the camera's motion, respectively:

$$\vec{e_x} = N(\vec{v5} - \vec{v4}),$$
$$\vec{e_y} = N(\vec{v4} + \vec{v5}),$$
$$\vec{e_z} = N(\vec{n}),$$
$$\vec{e_t} = dist \times N(\vec{p});$$

(4.4)

dist is the distance from the camera centre to the centre of the matrix code, $\vec{p}$ is a vector from the camera's centre to the centre of the matrix code on the image plane and N is a normalisation function.

The algorithm uses a camera as a visual tag reader (i.e. a barcode reader) to estimate the 3D position and orientation of a CyberCode symbol relative to the camera, and vice versa. With the visual tag recognition algorithm, the whole mechanism enables devices to be utilised as a barcode reader to retrieve data encoded in CyberCode symbols, to activate the associated actions and to attach information to the symbols.

The CyberCode system provides a wide variety of applications that bridge between the physical world and the digital world. One such example is the *3D annotation service*, an indoor navigation system in museums. CyberCode symbols are attached to every label that identifies each item in a museum. Visitors are provided with a browsing device called *NaviCam*, which serves as both a barcode reader and the datum point for estimating the distance between the reader and the real-world object, to which a CyberCode symbol is attached. The NaviCam is used to detect an ID tag (i.e. a CyberCode symbol) attached to an object, retrieve the corresponding 3D annotation information from a database, estimate the camera position and superimpose the retrieved information on the displayed video image. Figure 4.5 (left) gives an example of such a 3D annotation service.

With other sensor technologies, such as a gyro-enhanced ID recognition device, it is possible to provide a different type of navigation system that demonstrates a typical use of ID recognition with spatial awareness. A user first lets the device recognise the ID of the nearby CyberCode symbol; then, from the recognised ID and its shape, the system

3D annotation

3D navigation system

Fig. 4.5   Two different types of navigation system that use CyberCode tags (adapted from [7]).

determines its global position, including its orientation information, and continuously tracks its relative motion using its gyro-sensor. This enables the system to update its displayed navigation information correctly, thus allowing a user to look around an area without requiring the CyberCode symbol to be within the sight of the camera. Figure 4.5 (right) shows an example of a building-navigation application where a gyro-enhanced camera device is utilised. The FieldMouse system provides a similar application; the main difference is that the FieldMouse can only work on 2D surfaces whereas the CyberCode navigation application can provide a 3D AR environment.

CyberCode also can extend the physical concept of *drag-and-drop*, working as an operand for the direct manipulation of physical objects via a digital space. Such a drag-and-drop operation can be performed, for example, between a paper ID and an ID tag (i.e. a CyberCode symbol) attached to a printer, so as to invoke a printing job that produces a hard copy of the paper document. In Figure 4.6 (left) an *ID-aware pen* is used to perform a 'drag-and-drop' between the two physical objects, a paper document tagged with one CyberCode symbol and a printer tagged with another. With a hand-held barcode reader, this type of application can specify the source (i.e. the document) and destination (i.e. the printer) directly and can operate (i.e. print the document) regardless of network connectivity. Hence, this approach can be useful in the context of ubiquitous computing as well as in AR systems.

Hyper-dragging between a digital device and a physical object is another example of how a direct-manipulation technique can be extended into the physical world; see Figure 4.6 (right). A camera mounted above an augmented table and a PC with a CyberCode symbol are required for this application to work. The camera recognises the CyberCode or Internet Protocol (IP) address of the PC and the position on the table. By combining the CyberCode and its position information, the recognition system enables a seamless information exchange between the PC and the table. For example, a user can grab an object on the PC and drag it to the table by manipulating a cursor on the PC across the boundary of the computer.

To implement ubiquitously various AR applications in different situations and environments, a variety of image capturing and processing devices have been utilised and configured as tailor-made CyberCode readers to operate a particular application. Examples of such readers are the NaviCam, a gyro-enhanced camera device, an ID-aware pen

Drag-and-drop operation　　　Hyper-dragging between a PC and an augmented table

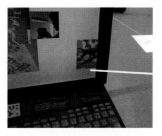

Fig. 4.6　　Examples of direct-manipulation operations using CyberCode tags (adapted from [7]).

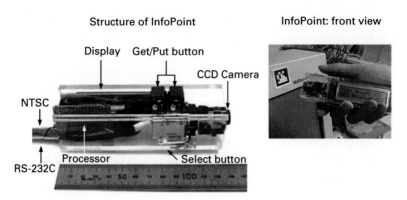

Structure of InfoPoint   InfoPoint: front view

Display   Get/Put button

CCD Camera

NTSC

Processor   Select button
RS-232C

Fig. 4.7   InfoPoint and its structure (adapted from [49]).

and cameras mounted above an augmented table or on a wall. An original hand-held direct manipulation device called InfoPoint (originally InfoStick) has also been developed[2] for the control of various digital appliances. InfoPoint is made up of a CCD camera for ID (i.e. barcode) recognition, buttons for operation and an LCD display for showing information about an object [7].

InfoPoint and its structure are presented in Figure 4.7. The functionality of InfoPoint is quite similar to that of the camera phones of today when they are used for barcode applications, except that InfoPoint does not provide wireless communication.

CyberCode was not invented specifically for mobile devices such as CCD camera-equipped mobile phones or PDAs. This is reasonable, considering that the first commercial camera phone manufactured by Sharp Co., J-SH04, was released by J-Phone[3] in 2000, the same year that the CyberCode project was introduced. However, the symbology is readable by the low-cost CMOS or CCD cameras found in most mobile devices today.

An aim in developing the CyberCode system was to promote the prevalence of 'augmented-reality-ready' mobile devices [7]. Although the system was not implemented on camera phones, this development highlighted the great potential for ubiquitous applications using 2D barcode technology and low-cost CMOS or CCD cameras. In fact, some CyberCode-based applications have been commercialised as a bundled software tool on the Sony notebook PC with a built-in camera [7]. The Sony game software PlayStation 3 also uses CyberCode.

The CyberCode symbology, its systems and various applications based on Cyber-Code have had a great impact on subsequent research in the field of AR and ubiquitous computing, especially that involving human–computer interaction. CyberCode itself has evolved, and applications that use Active CyberCode were introduced in 2006 [50]. The Active CyberCode system, whose aim is to improve CyberCode in terms of user friendliness, allows users to interact with CyberCode symbols using their extremities such as

---

[2] InfoStick was developed by Kohtake *et al.* [49].
[3] One of the mobile phone operators in Japan at the time, now known as Softbank Mobile Co.

Active CyberCode application example

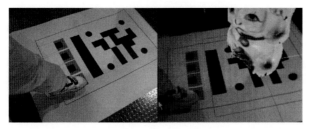

Active CyberCode variations

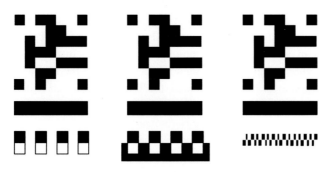

Active CyberCode   Active CyberCode with   Active CyberCode
                 enhanced robustness   with a slider

Fig. 4.8   An application example of Active CyberCode and its symbol variations (adapted from [50]).

fingers and feet to control the AR applications. The upper panel in Figure 4.8 presents an example where Active CyberCode is controlled by a user's foot and the lower panel shows some Active CyberCode variations.

Two years after the introduction of the CyberCode system, in 2002, J-Phone and Sharp Co. introduced the J-SH09, the first commercial camera phone with a reader for the 1D Japanese Article Number (JAN)[4] barcode and the 2D QR Code. This opened the door for ordinary users to interact actively with 1D and 2D barcode technologies in Japan.

4.2.2   Ubiquitous computing

In recent years the term 'ubiquitous computing' has come into its own. It was coined by Mark Weiser, as mentioned in [23]; he perceived personal computers at the time as so dominant that they colonised humans' lives, not just their desktops [46]. The goal of ubiquitous computing is to create an environment where the connectivity of devices is embedded in such a way that it is always available unobtrusively [46]. Some researchers have attempted to use 2D barcodes to make the ubiquitous computing world a reality.

---

[4] Japanese barcoding standard; JAN is compatible with the European Article Number and the Universal Product Code.

TRIP

The work in [20] introduced Target Recognition using Image Processing (TRIP), a low-cost and easily deployable vision-based sensor technology. In TRIP issues quite similar to those seen in CyberCode applications are addressed. A combination of 2D circular barcode tags or ringcodes (see Figure 3.24) and inexpensive CCD cameras such as webcam or CCTV cameras that plug into standard PCs is used to detect the identifier (i.e. the TRIP code) and the *pose* (i.e. the location and orientation) of a targeted object in real time with respect to the viewing camera in the system [20]. A distinctive feature of the TRIP code is its circular shape, which works better for the identification and 3D localisation of an object than other shapes (e.g. a square) when a visual sensor [20] is used.

Although there is a lot in common between the TRIP and CyberCode systems, the former is different from the latter in that the main focus of the CyberCode system is AR applications, whereas the goal of the TRIP system is to create a sentient computing (SC) environment, which is a form of ubiquitous computing. In contrast with the Cyber-Code system, where identifiers are often set statically and the relevant information is superimposed on the 3D location detected, the TRIP code is used to dynamically track the identifier attached to a moving object and to provide real-time corresponding 3D location information to make the SC environment a reality.

The aim of the SC system is to enhance the activities of users, providing them with the right service at the right time. The services are performed by computers and devices, embedded in the environment, which perceive their surroundings and become aware of the service that is required. The perception is conveyed by sensors that capture contextual information such as the location and identity of the objects in, or the ambient sound level and temperature of, a physical location. In the SC system, this perception is combined with static information such as an object's attributes, the geometric features of the physical location and the capabilities of the devices, in order to build a model of the state of the environment. The model helps the SC system determine the right service to be provided to an object at a particular time.

The TRIP code is used to obtain information about the object; hence, recognition and pose extraction of the target's TRIP code and its continuous tracking are necessary for the SC system to work. TRIP target recognition takes place in accordance with the following procedure.

(i) The image is binarised using an adaptive thresholding procedure. This operation removes the light effect from the reconstructed image and enables accurate detection of the regions of interest even when using low-cost, low-quality, cameras.

(ii) One-pixel-width edges are detected and thinned. The criterion for edge detection is that a pixel is taken to be an edge point if it has the intensity value of the colour black, an adjacent (i.e. four-connected) neighbouring pixel with the intensity value of the colour white and a diagonally adjacent (i.e. eight-connected) neighbouring pixel with the intensity value of the colour black.

(iii) The edges are followed and filtered. This is performed to keep only edges likely to define an elliptical shape, which could be perspective projections of the circular borders of a TRIP code.

(iv) A direct least-squares fitting of these ellipses, developed by [51], is applied. This operation obtains the ellipse parameters that best approximate the edge points.
 (v) The concentricity of the ellipse is tested in order to identify a candidate TRIP code.
(vi) The TRIP code is deciphered and validated against the two even-parity check bits reserved in the code [20].

Once the target object's TRIP code is detected, four pieces of information are retrieved from it. These are the ternary identifier, the parameters of the implicit equation of the outermost elliptical border of the projection of the TRIP code, the radius of the code and the image location of the bottom outermost corner of the synchronisation sector of the code [20].

Then, to extract the pose of the tagged object, a model-based object-location technique, namely POSE_FROM_TRIPTAG, is applied; this uses the outcome of the target recognition process as input to determine the accurate 3D location and orientation of the identified TRIP code with respect to the camera. This method establishes a transformation that back-projects the elliptical projection of the outermost border of the target into its actual circular form, of known radius, in the target plane [20].

Considering the high processing demand of vision-based sensors, which is often the main hindrance to their use, and the continuous TRIP operations, three different TRIP sensor operating modes are offered by the system according to the image processing needs for efficient and effective operation. These are the *default mode*, the *saving mode* and the *real-time mode*. As Figure 4.9 shows, the TRIP sensor switches its operating mode according to changes in the presence of the TRIP code in a given number of frames.

The TRIP default mode is activated either by a sensor start-up or whenever a timeout expires. The default mode switches to the TRIP saving mode whenever there are no TRIP code sightings in a given number of frames specified at the system start-up. Unless significant changes are observed, the sensor stays in this mode without consuming processing resources. Each time a TRIP code is spotted in the image, the default mode triggers the TRIP real-time mode. The timeout, specified as a parameter in the system's bootstrap, limits the time that the sensor spends in either the saving or real-time modes, where full-image analysis is not performed [20].

The TRIP system offers applications for human–computer interaction. *Active TRIP-board* and *TRIP Teleporting* are examples of such applications. By an interactive command issued by TRIP codes placed in the field of view of a camera, focused, say, on a whiteboard at a meeting, the Active TRIPboard allows the user to deliver the contents of the whiteboard as an instant handout to the attendees, either digitally (e.g. sending a Web link via email) or physically (e.g. hard copy of a snapshot of the whiteboard). The application components, namely, a *frame grabber* and a *TRIPparser* are activated either when a person appears in the camera view or through a Web interface. An *Active Badge indoor location system* is used to determine the presence of a person.

The TRIP-enabled teleporting service automatically activates or de-activates terminals of virtual network computing (VNC) according to the movement of its user. When a TRIP-code-wearing user comes within one metre of a webcam placed on top of a

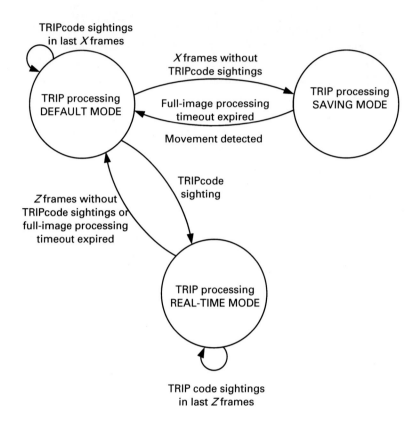

TRIPcode sightings
in last *X* frames

X frames without
TRIPcode sightings

TRIP processing
DEFAULT MODE

Full-image processing
timeout expired

TRIP processing
SAVING MODE

Movement detected

TRIPcode
sighting

Z frames without
TRIPcode sightings or
full-image processing
timeout expired

TRIP processing
REAL-TIME MODE

TRIP code sightings
in last *Z* frames

Fig. 4.9   The TRIP sensor operating modes and their triggers (adapted from [20]).

computer, the service starts monitoring and automatically displays the desktop associated with that user through VNC. This application is enabled by the fine-grained location sensor of the TRIP system, which is capable of measuring the precise location of objects in the physical world.

To manipulate and distribute the sensor data provided by the technology, an event-based distributed architecture was devised for the TRIP system [20]; this relies on a Common Object Request Broker Architecture (CORBA) infrastructure and a centralised recognition engine named 'TRIPparser'. To make it work, the system requires a full network with sufficient numbers of cameras and PCs in accordance with the scale of the SC environment.

In these studies, designing AR systems and/or ubiquitous computing environments has a higher priority than inventing better 2D barcodes or improving their capability. The use of barcodes is merely a means to implement the proposed applications. Accordingly, some researchers have claimed that traditional 1D barcodes are sufficient in terms of data capacity and may work better in certain applications. In fact, 1D barcode has been utilised in the FieldMouse applications, where barcode technology is used only to provide an ID of the starting point so that 2D locations based on it can be tracked.

Although 2D barcodes were invented for other approaches (i.e. CyberCode and TRIP code), the focus was then on designing AR and SC systems, respectively. In [7], for example, the effectiveness of using 1D barcodes as an alternative was argued, especially in applications that handle a large number of existing products. The role of 2D barcodes in the proposed applications for AR or SC systems is to provide IDs for the tagging systems. The capabilities, especially the data capacity, of these 2D barcodes are not as important as those of their counterparts in certain industries. It was claimed [7] that a large amount of data would no longer have to be encoded into ID tags in such an environment, where network-ready devices can provide seamless connectivity; in contrast, Yeh and Chen [44], as the inventors of a 2D barcode, put more emphasis on enhancing 2D barcode capability and appreciating that 2D barcodes are themselves portable databases rather than just a key to databases.

A highlight of barcode technology is its inexpensive operation. Similar systems and applications can be created using other technologies such as radio frequency identification (RFID) tags, near field communication (NFC) and infrared IDs. In fact, the significant progress in microelectronics and low-power semiconductor technologies has reduced the cost of RFID tags [22] over the years. However, when compared with barcode technology, whose cost is negligible, these other technologies are still very costly; this has hindered their widespread use.

Commonly available, inexpensive, PCs and CCD cameras are used to implement these barcode systems and the outcomes are rather promising, demonstrating the feasibility of the systems. However, in terms of creating ubiquitous computing environments these systems still have a common shortcoming: a lack of mobile devices that provide network connectivity. That is, these systems are only available in places where the required devices (e.g. cameras and PCs) are sufficiently integrated into the environment. If one has to be in a certain place to experience ubiquitous computing consciously, that is no longer considered to be truly ubiquitous computing [46].

To achieve true ubiquitous computing, network devices that are ubiquitously available or those with portability are required. Examples of such devices are mobile phones and PDAs. With the integration of built-in CCD cameras, and improved capability, mobile phones and PDAs have each become a type of miniature computer. They can perform multiple tasks, working not only as a networked device but also as a music player, a TV, a Web browser, an image capturing and processing device and so forth. As wireless mobile devices such as camera phones have gained popularity, they have become of interest to researchers.

4.2.3 Integration of barcode technology and wireless mobile devices

With the integration of enhanced capability, and its popularity, mobile phones have contributed a great deal to human lives, not only via human-to-human communication but also through human-to-technology interactions such as the short message service (SMS) and emails [23]. Consequently, they have established a position as a ubiquitous (also known as pervasive) computing tool. That is, mobile phones are within reach of their users most of the time and are thus available in many everyday situations [17], providing users with uninterrupted connectivity.

Effective and interactive shopping applications using 1D barcode

The paper [52] proposed a grocery shopping system called *iGrocer* where a smart phone is used as a scanner as well as a portable device that can provide wireless connectivity. The system helps users to perform a compatibility check between their health profiles and the nutrition content of a food item and to select healthy food, by providing information on each item on-line. It also enables queue-less secure checkout with the use of a credit card and real-time interaction with a map that shows the location of each product and the shortest path to it. Smart phones are used to display information obtained through the link to the relevant databases as needed and to perform queue-less checkout [52].

Although it addresses the requirements of both barcode technology and portable network devices such as smart phones with a scanner accessory, this approach is different from other studies of ubiquitous computing in that 1D barcodes are used in the system in the same way as they are used in grocery stores and supermarkets. In these other studies, barcode symbols are often used as visual entry points into the virtual world. That is, they work as a bridge to establish a connection between two different worlds, the physical world and the digital world. The data retrieved from the barcode symbol or tag may also be used as a source to perform additional operations.

The iGrocer system, however, merely provides a type of on-line shopping within the wireless network provided by a smart phone, without addressing the potential capability of barcode technology. The barcode symbols are used only to add items to the shopping list. This user-oriented application, however, proposing a new and efficient way of everyday shopping using barcode technology and a smart phone with scanning software, may direct users' attention towards the technology, resulting in its more widespread use.

Considering that 1D barcodes are printed on billions of product packages worldwide, if the proper infrastructures and systems are in place then there is a promising prospect of the widespread use of 1D barcodes for mobile applications [37]. Mobile devices with scanning capability will allow consumers to get access to a wealth of information about products, ranging from their descriptions, prices and reviews to price comparisons and the locations of retailers [53]. Such information can help them in their decision making and their selection of products.

In consequence, several 1D-barcode-reading algorithms have been proposed and developed that enable a mobile device to act as a robust and efficient barcode reader. While a mobile phone with a scanning accessory is used in the iGrocer system, mobile phones with a built-in camera that can function both as a communication device and an image capturing and processing device have been the target of recent researchers. The built-in CCD or CMOS camera of a mobile phone is used to perform a scanning task and, consequently, reading algorithms have been developed that take into account the limitations of these cameras.

Although most camera phones at the time (2004) had pre-installed barcode readers for EAN-13 and QR Code, Ohbuchi *et al.* in [54] proposed a new image recognition algorithm that uses a *spiral scanning method* for detecting the EAN-13 symbol. The algorithm is implemented as a mobile-application processor architecture made up of an embedded camera, a digital signal processor (DSP), a display device and so forth.

This algorithm consists of three components: pre-processing, detection of the black bars and sampling. After pre-processing, in which binarisation of the captured image takes place, the spiral searching method is applied from the centre of the image: parallel black bars are detected by scanning in a spiral pattern. Lines perpendicular to the black bars are sampled to calculate and determine the width of each bar and each space. This allows the algorithm to work accurately regardless of the orientation and any distortion of the captured symbol image. Based on the assumption that the centre of the captured image includes part of the symbol area, this algorithm reduces the computational cost of searching.

A vision-based technique proposed by [53] is another example of a 1D barcode decoding algorithm for camera phones and PDAs. This algorithm also comprises three main components, pre-processing, location of the target barcode and barcode decoding. In the pre-processing phase, luminance information is extracted first if the input image is not in a greyscale format. Then, depending on the size of the input image, the four edges are trimmed so that the image size becomes divisible by a block size of 32×32 pixels. This algorithm relies on both pixel-level and block-level processing [53].

In the second phase, the target barcode symbol is located in the following steps.

(i) The input image is divided into non-overlapping blocks; see Figure 4.10(a).
(ii) The greyscale data of each block are binarised using the *Otsu thresholding technique*; see Figure 4.10(b).
(iii) A morphological operation known as 'skeletonising' is performed on the binary data of each block; see Figure 4.10(c).
(iv) Each connected component within each block is extracted and labelled; see Figure 4.10(d).

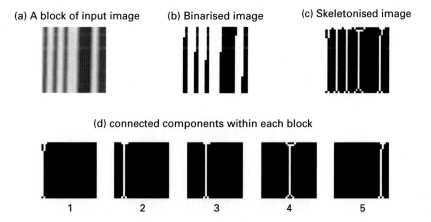

(a) A block of input image (b) Binarised image (c) Skeletonised image

(d) connected components within each block

1 2 3 4 5

Fig. 4.10 The processing of image examples (adapted from [53]).

(v)   The orientation of each labelled region within each block is calculated by measurement of the angle between the *x*-axis and the major axis of an ellipse that has the same second moments as the region.

(vi) The orientations calculated to test for the parallel property are compared.

(vii) The barcode region indicated by a group of blocks with similar parallel line patterns is extracted [53].

Once the target symbol has been located, the focus of the third phase is to decode it. First of all, a sample row across the barcode image is obtained (see Figure 4.11(a)). This is then converted into binary form by simple thresholding based on the mean value of the signal (see Figure 4.11(b)). The signal in binary form is essentially a binary sequence of '1's and '0's. To determine the individual widths of the alternating black and white bars, *run-length encoding* is performed on the binary sequence.

When reading EAN-13, the first step is to locate the left-hand guard pattern, which indicates where to start decoding the symbol. Then, the next six sets of four bars are decoded into six digits (i.e. from the second to the seventh digit) before the decoder reaches the centre guard pattern. Another six sets of four bars are then decoded into six digits (i.e. from the eighth to the 13th digit) until the decoder reaches the right-hand guard pattern, which signifies the end of the barcode. The result obtained for the example symbol (see Figure 4.11(b)) is 315693 510776 (see Figure 4.12(a)). The parity pattern on the left-hand side of the barcode determines the first digit. The parity pattern OEEOEO,

(a) Sample row across the barcode image

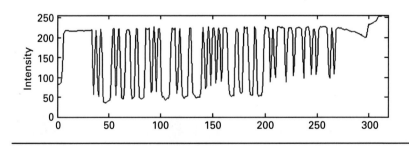

(b) Thresholded sample row

Fig. 4.11   (a) A sample row across the barcode image and (b) its binary form (adapted from [53]).

(a)

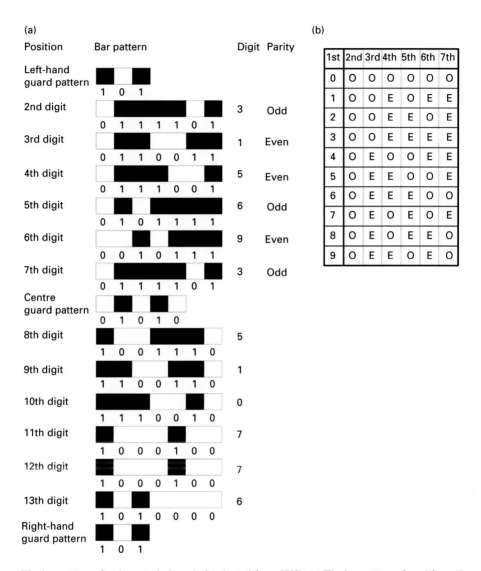

(b)

	1st	2nd	3rd	4th	5th	6th	7th
0	O	O	O	O	O	O	O
1	O	O	E	O	E	E	
2	O	O	E	E	O	E	
3	O	O	E	E	E	E	
4	O	E	O	O	E	E	
5	O	E	E	O	O	E	
6	O	E	E	E	O	O	
7	O	E	O	E	O	E	
8	O	E	O	E	E	O	
9	O	E	E	O	E	O	

Fig. 4.12   The bar patterns for the sampled symbol (adapted from [53]). (a) The bar patterns found from the threshold sample row. (b) The first digit and its parity pattern: O, odd; E, even. When the first digit is 0, we have a UPC code.

where O and E denote odd and even, respectively (see Figure 4.12(b)), specifies that the first digit is 9, retrieving 9 315693 510776 all together.

Once all 13 digits are obtained, the last step is an error check to verify the correct reading of the symbol. The algorithm involves multiplying each of the first 12 digits of the barcode with a weight and then summing the results together:

$$weighted\_sum = \sum_{i=1}^{12} weight(i) \times digit(i), \qquad (4.5)$$

where the weight vector is [1 3 1 3 1 3 1 3 1 3 1 3]. The checksum is the smallest number which, when added to weighted_sum, produces a multiple of 10. In this example, the weighted_sum is $9 \times 1 + 3 \times 3 + 1 \times 1 + 5 \times 3 + 6 \times 1 + 9 \times 3 + 3 \times 1 + 5 \times 3 + 1 \times 1 + 0 \times 3 + 7 \times 1 + 7 \times 3 = 114$. Hence, the checksum digit is 6, since when it is added to the weighted_sum the result is 120, which is a multiple of 10. The checksum matches the 13th digit of the barcode, which validates the barcode value produced by the decoder.

The applications for 1D barcode technology have already been well established in various fields. The technology itself as well as its markets have reached their maturity. Furthermore, standardisation of the symbologies was essential to allow the entire 1D barcode system to work effectively. Such a background limits diversity in developing mobile applications that can use 1D barcode. Although the idea of using mobile devices for reading 1D barcode symbols and making them available for consumers might be innovative, the novelty and variation in applications that use 1D barcode are dwarfed by those developed for its 2D counterpart.

A variety of cutting-edge applications have been developed based on the use of 2D barcode technology with wireless devices, especially camera phones. Targeting these two mobile technologies, some researchers have devised user-oriented applications of the existing 2D barcodes. Some have invented their own 2D barcodes to implement their novel mobile applications. In contrast with earlier research, where systems were designed for indoor scenarios, these 2D barcodes and applications were developed for use in both indoor and outdoor environments. The proliferation of mobile devices has enabled communications not only between humans but also between humans and electronic devices, regardless of time and location.

A wide variety of mobile applications using 2D barcodes
QR Code

Quick Response Code (see Section 3.1), invented by Denso Wave in 1994, is one of the earliest 2D barcodes utilised in applications where a combination of camera phones and 2D barcode technology is required [24]. The use of QR Code in this way was not the original intention of the inventors. However, technologies often flourish when they match the needs of the time, regardless of the intentions of the inventors.

Since commercial camera phones with a QR Code reader became available in 2002, a wide variety of QR Code applications for mobile devices have been devised by not only mobile-content-providing companies but also general users of applications in Japan, resulting in their widespread use. As QR Code applications have proliferated, the number of companies that provide users with a free Web-based generator and decoder of QR Code has increased, which, in turn, has facilitated the development of user-oriented QR Code applications. These free readers and generators allow users not only to decode the QR Code symbols but also to create their own QR Code symbols. These symbols can then be exchanged over the Internet, saved on mobile phones or printed.

Meanwhile, the focus of researchers has been on the development of algorithms to improve the reading robustness of QR Code symbols rather than on developing new applications. In [54], for example, a new recognition algorithm for QR Code was proposed that detects the fourth corner of a symbol using the positions of the finder pattern located at the other three corners of the symbol. In the existing algorithm, the detection and projective transformation were performed using the three corners of the finder pattern along with the alignment pattern, whereas the new algorithm uses the four corners of a square symbol to carry out both tasks. Hence, the proposed algorithm can be applied to other square symbologies.

As the reading robustness of the QR Code is being improved, and mobile devices have become more sophisticated, new types of QR Code have been created. Examples are coloured QR Codes, design QR Codes and animated QR Codes. Provided that the contrast between the two colours used is sufficient to be distinguished by the code reader, the coloured QR Code symbols can be robustly decoded. The strong error correcting capability in QR Code also allows a small illustration and logo to be integrated into its symbol. When QR Code symbols are displayed on the PC screen, the animated images in the background of, or within, a QR Code symbol are quite eye catching. These design QR Codes are often developed and used for commercial purposes.

Currently, QR Code symbols are seen everywhere in Japan. The readers for QR Code as well as 1D barcode are pre-installed in nearly all Japanese camera phones as one of their default functions. The symbols are displayed on Web sites, in magazines, newspapers and posters. A blog written in QR Code symbols has even appeared. This code, which works as a portable database, can be used when no network connectivity is available. For example, it allows users to store personal details and contact information, encoded in QR Code symbols, in their phone address books by simply capturing the QR Code symbol printed on a business card. The captured QR Code symbol is decoded on the phone itself, ready for saving. The symbols also work as portable files, providing e-tickets and e-coupons. For example, Japanese airlines offer a self check-in service using QR Code e-tickets.

Recently, QR Code applications that are more relevant to a consumer's everyday life have become popular. For example, in responding to growing concerns about produce safety, agricultural organisations have implemented food-history information services as part of projects to improve produce quality and to assure consumers of food safety by providing the details of their produce. The code works as a medium for tracking, a portable file that carries produce information and a tag to provide the link between producers and consumers. By capturing the QR Code on the package, consumers can access information on the produce, including the supplier's contact details, harvest date, shipment date, fertiliser and agricultural chemicals used [37].

Among all these applications, the most popular QR Code application is that linking a camera phone to the Internet, where users have uninterrupted access to online activities such as online shopping and checking real-time information on public transport and events. This trend has been observed for most 2D barcodes, especially index 2D barcodes for mobile applications. In 2006, as a part of Windows Live service, Microsoft launched the Windows Live Barcode, where QR Code is used to bridge information between mobile devices and other media such as PCs, plasma screens and magazines. It was followed by the release of Windows Live Confucius in 2007. Windows Live Confucius provides users with both online and offline QR Code generators, allowing them to be active users of QR Code. Such activities may facilitate an even greater proliferation of QR Code applications beyond Japan.

Semacode

Semacode is one of the earliest database 2D barcodes specifically targeting camera phone users. Semacode is, in fact, a trading name of the Data Matrix symbology. Semacode Co., a Canadian firm founded by Simon Woodside, has utilised the public-domain standard Data Matrix 2D barcode format basically to encode plain-text uniform resource locators (URLs) [30] and to provide hyperlinks between the physical and digital worlds. By using camera phones that support Semacode applications, URLs are automatically read out of a Semacode symbol and presented in the phone browser [30].

Examples of novel Semacode applications are up-to-the-minute information about bus services and virtual treasure hunts [55]. In the former application, a Web page accessed via Semacode provides, for example, the arrival time of the next bus. The latter application turns our everyday physical world into the stage for a virtual game. Participants hunt for Semacode symbols embedded in the physical world. When the 'treasures' or Semacode symbols are found, the participants can claim their prizes by simply capturing the symbols with their camera phones. The participants may find their prizes from either the textual data directly encoded in the Semacode symbol or the Web page to which they were directed by the URL encoded in the symbol.

When Semacode and its applications were first introduced, using 2D barcode and a camera phone to connect the physical and digital worlds was not as common as it is now. Along with codes like QR Code and Data Matrix, Semacode has played an important role in the gradual spread of such technologies and their applications. In this respect, Semacode is considered to be one of the pioneering 2D barcodes used for mobile applications.

As was described in Chapter 3, Data Matrix is itself a robust 2D barcode that can store hundreds of characters (the maximum theoretical density is 500 million characters to the inch) and can recover up to 50% of data lost owing to damage [28]. Semacode has demonstrated its capability and usability for mobile applications, leading other companies to adopt the Data Matrix format to implement their own mobile applications. UpCode, introduced by the Finnish company, UpCode Ltd., is an example.

In some approaches, existing barcode symbologies are used to develop applications that target general users of camera phones; QR Code, Semacode and UpCode are examples of such approaches. Conversely, some researchers and developers have adapted existing 2D barcodes that suit the applications they have devised. For example, Madhavapeddy *et al.* [34] adapted TRIP code to operate it with mobile devices in wireless environments. The resulting derivative of TRIP code was called SpotCode. Another example is Visual tag, which derives from both TRIP and SpotCode [35].

Human–computer interaction via 2D barcodes
SpotCode

Whereas the purpose of developing mobile applications for QR Code, Semacode and UpCode was to provide users with convenient and flexible ways to perform commercial

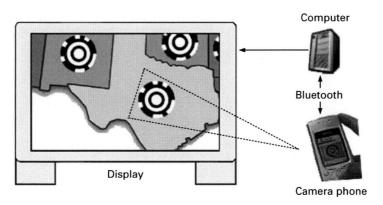

Fig. 4.13   Components of the SpotCode system and the data flow between components (adapted from [34]).

activities using URL links established via these 2D barcodes, the interest of academic researchers has been to pursue a better way of establishing ubiquitous computing environments. As a solution, SpotCode was designed specifically to enable human–computer interaction, using camera phones and 2D barcode technology as a visual tag. The mobility of camera phones as image processing devices enables place-independent applications.

Three major components, namely, a Bluetooth-enabled camera phone, a Bluetooth-enabled computer and SpotCode symbols, are required for the SpotCode system to operate. Both *active* SpotCode (e.g. SpotCode symbols displayed on a computer screen) and *passive* ones (e.g. SpotCode symbols printed on paper) work robustly in the system. Figure 4.13 illustrates the components required for the system and the data flow between them.

A SpotCode contains two pieces of information: a *service identifier* and a *data block*. The service identifier informs the camera phone of the relevant Bluetooth service to a certain SpotCode and the data block consists of application-specific information [34]. While the camera phone is used to decode the information, the short-range wireless communication provided by the Bluetooth services enables data transmission between the SpotCode and the camera phone. In addition to the three core elements described above, a physical medium or active/passive display on which SpotCode symbols are placed is necessary for the SpotCode system.

Through the established wireless link the phone constantly transmits the data block of the codes along with the current (x, y) positions, symbol sizes and orientation of the SpotCode symbol relative to the position of the camera [34]. As a result, the system allows a Bluetooth-enabled camera phone, acting as a sophisticated pointing-device, to control active displays. Examples of two different 'SpotCode jukebox' controllers are presented in Figure 4.14. Users can control the jukebox's volume by rotating their camera phones. The amount of rotation is indicated by a triangle indicator above the SpotCode symbol, and changes in the volume can be observed by both the percentage displayed above the symbol and the sound indicator next to the symbol. By moving a

Rotation control for volume change   Slider control for track selection

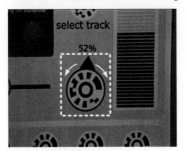
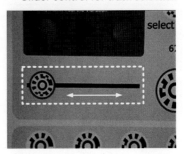

Fig. 4.14   Examples of SpotCode jukebox controllers (adapted from the Computer Laboratory SpotCode
Interfaces Web site of the University of Cambridge).

camera phone right to left and thus sliding the SpotCode symbol forward and backward,
users can easily select, move or skip a location on the music track.

SpotCode also works as a user interface for devices without a display or input capability
of their own, using the mobile phone's display and keypad as the point of human–
computer interaction [34].

Bluetooth device discovery

Madhavapeddy and coworkers have explored further the usability of visual tags in a
Bluetooth-based wireless communication environment. The focal point of their new
study was on how effectively visual tags can be used to bypass the Bluetooth device-
discovery process. According to their work reported in [35], despite the proliferation of
mobile devices with Bluetooth capability, the use of Bluetooth as a networking technology for publicly accessible mobile services has been limited owing to its inappropriate
and ineffective way of device discovery. By implementing an end-to-end Bluetooth-
based mobile service framework, these workers demonstrated the effectiveness of a
tag-based connection-establishment technique, developed over the standard Bluetooth
device-discovery model, in terms of speed, usability and scalability.

For the purposes of their research, the circular tags used in the Bluetooth device-
discovery system, initially inspired by TRIP codes, were improved in two ways, as
follows:

(i) the number of bits encoded in a single Visual tag symbol was increased by the adding
of *extra data rings* (four rings were used, adding two extra rings to the SpotCode
symbology); and

(ii) an explicit *sync marker* was added to specify the orientation of the Visual tag symbol.

Figure 4.15 presents the TRIP code, SpotCode and Visual tag symbols used for
improvement of the Bluetooth device-discovery system.

The image processing technique developed for SpotCode reading was also integrated
into the proposed system. Each barcode contains a 48-bit Bluetooth Device Address
(BD_ADDR) and 15-bit application-specific data.

TRIP code   SpotCode   Visual tag

Fig. 4.15   TRIP code, SpotCode and Visual tag (adapted from [35]).

In the tag-based device-discovery system, a device being detected must have a tag (i.e. a barcode symbol) attached to it, either in a physical print format or as an image displayed on the device's screen (e.g. as the default wallpaper). To establish Bluetooth connection between two mobile phones, for example, one user is required to aim his or her camera phone at the tag attached to another device with which he or she wishes to communicate. Once the image is captured by the camera phone, the BD_ADDR of the phone being detected is retrieved and a connection between the two devices is immediately established.

In the case where the 15-bit application-specific data are encoded in the tag together with the device's address, the associated service is also provided. Hence the system provides not only faster Bluetooth connection but also easier service selection than the standard Bluetooth device-discovery model, where a user is required first to select the device's name with which he or she wishes to communicate among all the listed device names and then to make a service selection among all available services. The system works efficiently, especially when the user is aware of the physical location of the device with which he or she wishes to establish a connection.

SiD: human-verifiable authentication system

A mechanism similar to the Bluetooth device-discovery system described above is used in the seeing-is-believing (SiB) authentication system proposed in [56]. Generally it is rather difficult to ensure which device is at the end of a wireless connection, and thus wireless communication is often considered to be vulnerable against malicious attacks such as a man-in-the-middle (MITM) attack. In the SiB system, the combination of barcode technology and camera phones is used as a visual channel for human-verifiable authentication. In order to implement authentication between two camera phones, say, phones A and B, using the SiB protocol, the following procedures take place.

(i) Each phone encrypts a cipher text of their public key and generates a 2D barcode encoding this cipher text.

(ii) Phones A and B take turns displaying and taking snapshots of their respective 2D barcodes (the *pre-authentication* phase). Both devices now hold a cipher text of the other device's public key.

(iii) Phones A and B exchange their public keys over the wireless communication link.

(iv) Each device performs the same encryption function using the other device's public key, ensuring that the result matches the cipher text that was received over the visual channel (the *mutual authentication* phase) [56].

This authentication procedure is conducted visually in the presence of the users and their devices and thus each can be sure that his or her device is communicating with the intended device (*demonstrative identification*). Only tangible visual tags such as 2D barcodes enable such strong authentication. This research is important in that it has demonstrated the potential use of the integration between 2D barcode technology and mobile phones in a variety of contexts, including in the establishment of secure wireless communication.

Visual Code

The SiB system adopted the Visual Code symbology, reported in [17]. The most important aspect of Visual Code is that it was developed explicitly for use with camera-equipped mobile phones. In order to ensure the robustness of the Visual Code system, the limitations inherent in the built-in cameras of mobile phones and the mobility of such cameras are specifically addressed in the development of the symbology itself, its system and applications.

Generally, the quality of the built-in cameras in mobile phones is lower than that of digital cameras. Mobile phones usually offer lower camera resolutions. They often lack functions such as auto-focus and zooming. Furthermore, the capability of a mobile phone as a computing device is rather limited. For example, mobile phones often lack a floating point unit (FPU). When the Visual Code system was developed, the difference in capability between camera phones and digital cameras was greater than it is today. These camera phones cannot be compared with dedicated barcode scanners. Hence, it was necessary to address these operation limitations in order to implement a robust visual tagging system.

Owing to the mobility inherent in camera phones, the symbols scanned by a camera phone may appear at any orientation in the captured image. Symbol distortion is also introduced in these images. These factors complicate the process of image recognition. However, such apparent shortcomings have led to proposals for some novel features in the Visual Code system. In [17], Rohs conceived an application that can provide different information according to the amount of tilting and rotation of a symbol, by continuous tracking and measurement of the symbol.

To implement such an application, Rohs proposed a code coordinate system, visual detection of the mobile phone movement and determination of the rotation angle and amount of phone tilting. The integration of these features allows users to obtain multiple information items located at known code coordinate positions relative to a single Visual Code symbol [17]. It also changes a camera phone into an optical mouse or a remote controller that enables ubiquitous interaction between a human and a computer.

According to [17], the quality of the colour image generated by the built-in cameras on mobile phones was rather poor when the Visual Code symbology was developed. The resolution of most built-in cameras at the time was video graphics array (VGA, i.e. 640 × 480 pixels) or less. This feature determined the minimal cell size of the symbology, which is rather large in comparison with most 2D barcodes developed for industrial use. Another reason why the minimal cell size is rather large is that the applications developed by [17] were to extend the visual tagging system, where data capacity is not a first priority.

As previously discussed, the goal of visual tagging systems is to realise AR and/or ubiquitous computing. In order to interact with computing devices such as a large-scale wall monitor, which could be located either inside or outside a building, under varying light conditions, a barcode should be sufficiently large and distinctive to be recognised from a distance. This is also a reason for using a black and white format in the symbology. This format can produce a clearer contrast than the use of any other two colours [17].

4.3 Development of colour 2D barcodes

4.3.1 Using colour elements for improving visual appearance

The use of colour in a barcode can be advantageous in two respects. Coloured images are generally more visually appealing than their monochromatic counterparts. This is highly appreciated especially when 2D barcode is used for advertising purposes. Another advantage of using colour is that it can improve data capacity without increasing the number of cells in a barcode symbol. A colour cell can represent more than one bit, depending on the number of component colours used for encoding the data. When four different colours are used to encode data, one colour cell can represent two bits. When eight different colours are used, three bits can be encoded in a cell.

However, there is a trade-off between the number of colours used to encode the data and the reading robustness of the barcode. The more colours are used in a colour space, the smaller the distances between neighbouring colours will become, resulting in less contrast between the different colours. A low contrast between colours indicates that the values of the colours are close, which makes it more difficult to distinguish the colours and, consequently, to retrieve the original data.

There are other factors that should be addressed when using colour symbols. First of all, it is more expensive to use colour images both computationally and in terms of their physical production. It is considered that producing and processing colour images is expensive, as they require more sophisticated devices. The second factor is the availability of colour printers. Generally, monochromatic printers are more widely available. The last and most critical factor is the difficulty in reproducing the original colour values after the required image processing operations have been performed on the images. Various factors may change the values of the original colours during the image processing operations. These include the lighting conditions, the image processing devices and the media on which the colour images are displayed or printed. Time is also a factor since colours often fade away as time passes. As a consequence, researchers often prefer to develop black and white barcode symbologies rather than colour symbologies.

However, in [40] Han *et al.* demonstrated the effective use of reference colours to overcome these problems in the retrieval of the original colour values. The ColorCode invented by Han and his team at Yonsei University uses reference cells that provide a standard colour for correctly distinguishing each reproduced cell colour. The value of each cell colour in the data area is determined relative to the value of the standard colour in the reference cells. Because the relative difference between the cell colour and the standard colour is consistent, a reader can correctly retrieve the original data.

ColorCode is designed to work with any devices with a CCD or CMOS camera and an Internet connection. Examples of such devices are camera phones and a PC with a webcam. This allows the ColorCode technology to be used ubiquitously, which led to the implementation of the Ubiquitous-computing Town Project: Intelligent context Awareness (UTOPIA) project. The UTOPIA project aimed to develop user-centred services provided in context-aware, ubiquitous-computing, environments [57]. The project had three stages, namely, U-Campus, U-Town and U-City, expanding its boundary as the stages progressed.

For example, the U-Campus services include U-Messaging and U-Campus Tour Guide services targeting students, staff and visitors of a university [57], whereas the U-Town and U-City projects provide town-wide and city-wide services such as the U-Restaurant service, the U-Museum service and the U-Theme Park service [57]. Owing to its inexpensive operation and readiness for commercial use, ColorCode symbols were used as visual tags to bridge the physical and digital worlds for the UTOPIA project. Mobile devices such as camera phones were used to read the ColorCode symbols and establish network connections to provide users with designated services.

ColorCode was not the first colour 2D barcode to be introduced. For example, Ultracode, developed by Zebra Technologies uses both monochrome and colour symbols to encode data. However, ColorCode was the first colour 2D barcode developed for use with low-resolution, inexpensive, CCD cameras such as webcams, and the built-in cameras of mobile phones. Its invention and successful implementation has demonstrated both the technological and economical feasibility of systems where coloured 2D barcode is used as a visual tag to bridge the physical and digital worlds. It has had a considerable impact on subsequent research pursuing effective use of the colour element in new 2D barcodes.

Furthermore, ColorCode is one of the pioneering 2D barcodes that have given themselves additional value by integrating graphical elements into a symbol. Being an index 2D barcode whose cell size is rather large, ColorCode decoding requires only 40% visibility, i.e. the actual data need to occupy only 40% of each individual cell. This allows a ColorCode symbol to be incorporated into graphic designs, making the most of the remaining 60% of the space. Thus, a ColorCode symbol can perform two tasks simultaneously: as a visual tag and as an advertisement. This trend has also been followed by many new 2D barcodes such as mCode, BeeTagg and Trillcode.

4.3.2   Using colour elements for increasing data capacity

Despite its potential for increased data capacity, ColorCode has presented itself as an index 2D barcode that encodes only 10 digits. ColorCode uses colour elements for improving the visual appearance of the symbol. Moreover, another well-known coloured 2D barcode, the High Capacity Color Barcode (HCCB) developed by Gavin [42], uses colour elements to increase the data capacity of a 2D barcode occupying a given space.

This barcode was initially developed to encode biometric IDs, medical insurance or information cards, which often require a considerable amount of data capacity because they contain personal information such as an iris scan, a fingerprint image, a signature image, medical history or DNA profile.

In order to improve the data capacity for a given symbol space, HCCB combines two encoding schemes, one using colour elements and the other using triangular cells, to encode data (see Figure 3.33). The former allows more than a single bit to be encoded in a cell whereas the latter enables more cells within a given space since triangular cells require less physical space than square cells. Using these schemes, an HCCB symbol can achieve at least three times the density of industrial-standard 2D barcodes such as the PDF417 and the Data Matrix [42].

With a black and white encoding scheme, each colour represents only a single bit, either 1 or 0. In contrast, it is possible to encode more than one bit within a single cell by using a colour encoding scheme, since the number of bits encoded in a single colour cell depends on the total number of colours used for encoding data. Generally, the more colours that are used, the more bits can be encoded within a single cell.

However, as mentioned earlier, using more colours makes it difficult to discriminate colour values. For example, if eight colours are used to encode data, these eight colours must be clearly discriminated to retrieve the original data. Monochromatic 2D barcodes normally use the two colours that are furthest apart within a colour space, namely, black and white. Since the distance between the two colours is a maximum, each colour can be robustly distinguished even when the size of each cell is small. This also enables robust colour discrimination in the poor images produced by low-cost and/or low-resolution devices such as built-in cameras in mobile phones.

There are two factors that aggravate the negative effect caused by using more colours: the size of each cell and the fidelity of the reconstructed symbol image. When the cell size is small, it is likely that the entire cell will be affected by the surrounding colours. The fidelity of the reconstructed images produced by low-resolution devices is often degraded by blurring artefacts at the edges between cells. To minimise such negative effects and increase reading robustness, HCCB uses *white space separators* between the colour cells encoded in a symbol. According to [42], triangles are less prone to blurring artefacts (also known as anti-aliasing effects) than squares because they present only three straight-lined sides. Hence, the triangular cells of HCCB also help minimise blurring artefacts. Like ColorCode, HCCB uses reference colour cells, called a reference palette, to retrieve the original data successfully.

Paper Memory Code (PM Code), developed by Content Idea of ASIA Co., is another colour 2D barcode that aims to improve data capacity by using colour elements. A unique feature of PM Code is that it uses a layered structure. A PM Code symbol is made up of a *surface layer* and a plurality of *code layers*. Each code layer is composed of a 2D matrix of colour cells and spaces (e.g. red cells and spaces, green cells and spaces, blue cells and spaces). To unify the code layers, the surface layer includes an *index information code* that contains the RGB (red, green, blue) values for each code layer; furthermore, the surface layer presents the resultant additive colour of all the code layers. Figure 4.16 shows an example of a PM Code symbol.

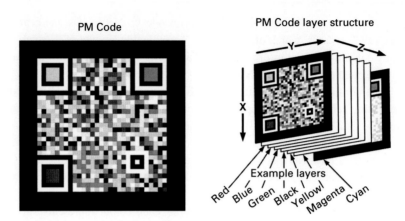

Fig. 4.16   Paper Memory (PM) Code symbol and its layered structure. For a colour version, please see Plate 4.

Each cell in the surface layer may present the colour of a single code layer or the resultant colour from adding the plurality of code layers. When the resultant colour is identical to the colour used in one of the layers, this resultant colour will be converted to a designated colour according to the PM Code colour-conversion algorithm. Two colour spaces are involved in this algorithm: the RGB colour space and the HSB (hue, saturation, brightness) colour space.

The RGB colour space, which stems from the three primary colours of light, is an additive colour space where a wide range of colours are produced by adding various combinations of the primary colours. It can be represented as a three-dimensional (3D) cube with red, green and blue located at three corners (see Figure 4.17(a)). A detailed description of the RGB colour space is provided in Section 6.2.2.

The HSB colour space is also known as HSV (hue, saturation, brightness or value). *Hue* refers to a basic colour, e.g. blue or cyan; *saturation* refers to the grey content of the colour; *brightness* or *value* refers to the intensity of the colour. This colour space is often used by people who select colours from a colour wheel or palette since this colour system is closer than the RGB system to the way in which humans experience and describe colour sensations [58]. The HSB colour space can be obtained by observing the RGB colour cube along its greyscale axis. The result is a hexagonal colour palette (see Figure 4.17(b)) [58]. The greyscale, shades of grey produced by equal values of red, green and blue, forms a diagonal line joining the black and white points. In the RGB colour space, black is at the origin and white is at the diagonally opposite end of the colour cube, which correspond to the black and white points in the HSB colour space.

During the PM Code colour conversion process, first the RGB values of the resultant colour are converted to the HSB values according to the PM Code colour-conversion algorithm. Then, by an inverse operation, the HSB values are converted to their corresponding RGB values [59]. The PM Code colour-conversion algorithm, together with its index information code, enables the decoding software to detect the presence or absence

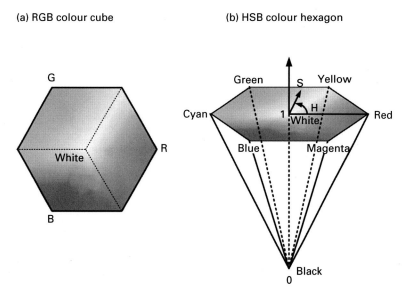

(a) RGB colour cube

(b) HSB colour hexagon

Fig. 4.17 (a) Red, green, blue (RGB) colour cube and (b) hue, saturation, brightness (HSB) colour hexagon cone, defined by red, yellow, green, cyan, blue and magenta axes radiating from the origin (black). For a colour version, please see Plate 5.

of colour cells in each layer. This enables the successful reading of each code layer (see Figure 4.16), resulting in the decoding of the entire PM Code symbol. Figure 4.18 demonstrates how the colour of each cell on the surface layer (see above) is determined from the PM Code colour-conversion algorithm.

All the examples (i.e. (a)–(e)) in Figure 4.18 should be printed in yellow, since the maximum value of each of R, G and B in the RGB colour space in the range [0, 255][5] is 255. However, the combination of the code layers in each example is different. In example (a), a single layer of yellow (RGB = 255, 255, 0) is used to compose the surface layer. However, two code layers, namely, green and red, yellow and red, and green and yellow, are used to compose the corresponding surface layer in examples (b), (c) and (d), respectively. The corresponding sums of the RGB values for (b), (c) and (d) are, respectively: 255, 255, 0; 510, 255, 0; and 255, 510, 0. In example (e), three code layers, namely, green, yellow and red, compose the surface layer, resulting in the summed values 510, 510, 0.

In the RGB colour space, R, G and B values that are more than the maximum value (i.e. 255) are neither displayed nor printed. Hence, PM Code uses the HSB colour space to differentiate the resultant colour of each surface cell by the colours of its component code layers. The summed RGB values are first converted to HSB values according to the colour conversion algorithm and then converted back to the RGB values of the corresponding colour by an inverse operation [59]. This algorithm enables successful

---

[5] For an eight-bit resolution image, there are 256 possible levels for each R, G and B component. This gives a range from 0 to 255.

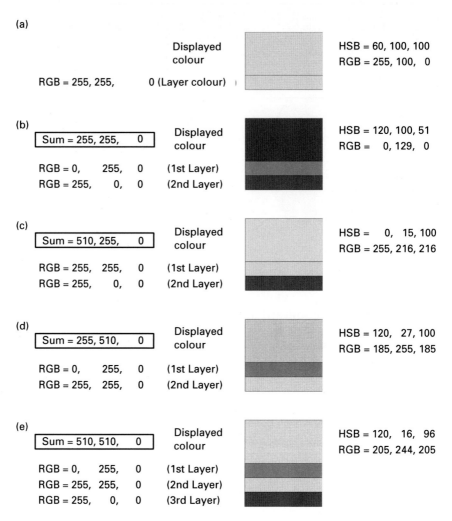

(a)

Displayed colour   HSB = 60, 100, 100
                     RGB = 255, 100, 0

RGB = 255, 255,   0 (Layer colour)

(b)

Sum = 255, 255,   0   Displayed colour   HSB = 120, 100, 51
                                               RGB =   0, 129, 0

RGB = 0,   255, 0   (1st Layer)
RGB = 255,   0, 0   (2nd Layer)

(c)

Sum = 510, 255,   0   Displayed colour   HSB =   0, 15, 100
                                               RGB = 255, 216, 216

RGB = 255,   255, 0   (1st Layer)
RGB = 255,   0, 0   (2nd Layer)

(d)

Sum = 255, 510,   0   Displayed colour   HSB = 120, 27, 100
                                               RGB = 185, 255, 185

RGB = 0,   255, 0   (1st Layer)
RGB = 255,   255, 0   (2nd Layer)

(e)

Sum = 510, 510,   0   Displayed colour   HSB = 120, 16, 96
                                               RGB = 205, 244, 205

RGB = 0,   255, 0   (1st Layer)
RGB = 255,   255, 0   (2nd Layer)
RGB = 255,   0, 0   (3rd Layer)

Fig. 4.18   PM Code colour conversion (adapted from [59]); see also Figure 4.16. For a colour version, please see Plate 6.

retrieval of the colour values in the code layers that compose a PM Code symbol, resulting in the successful reading of the PM Code. However, in order for the algorithm to operate robustly the reading software must have a refined colour recognition capability since it must distinguish a wide range of colour values.

Neither HCCB nor PM Code were specifically developed for use with mobile devices; they are also used with other devices such as PCs. The data capacities of both these colour 2D barcodes are greater than those of their monochromatic counterparts when an identical number of cells are encoded within a given symbol space. This indicates that, in scenarios where dedicated scanners or their equivalents are used to decode the symbols, these colour encoding schemes can significantly improve the data capacity of a 2D barcode symbol.

However, when a resource-limited mobile device such as a camera phone is used as an image processing device, the minimum cell sizes for monochromatic and coloured 2D barcodes could be different. The former can be more robustly detected than the latter even when the cell size is small. That is, monochromatic 2D barcodes are capable of encoding a greater number of cells for a given space. Hence, increasing the number of colours to encode data does not directly lead to an increase in the data capacity of a 2D barcode symbol, especially when mobile devices are used.

Consequently, a careful selection of the colours for encoding data and a robust colour detection scheme must be made in order to create a true colour database 2D barcode optimised for mobile applications.

4.4 Summary

Barcode technology, especially for 2D barcodes, is considerably innovative. The trend observed in early studies on this subject is similar to those seen in most emerging industries and technologies. In the beginning, researchers attempted to either enhance the capability of 2D barcodes or develop applications that exploited their potential, especially their high data capacity.

Later on, when it was both technologically and economically feasible, the concern of researchers evolved to creating visual tagging systems using barcode technology, especially that involving 2D barcodes. Their goals were to realise augmented reality systems and/or ubiquitous computing. In order to do so, a fully networked environment was thought necessary in the early days of developing such systems.

Inspired by the emergence of camera phones, researchers have sought to explore the potential for integrating these two new technologies, namely, 2D barcodes and camera-equipped mobile phones, through the development of novel mobile applications. Examples of such applications are the Visual Code system and the SpotCode system, where 2D barcode technology, together with wireless communication devices, is used as an interface to bridge the physical and the digital worlds, enhancing human–computer interactions. These studies have demonstrated the feasibility of these novel systems.

The successful implementation of this research has led companies to develop systems that provide users with convenient and flexible services based on Web links established via 2D barcodes and mobile devices such as camera phones and PDAs. The mobile devices perform the image processing tasks while providing continuous Internet connectivity. As a result, a variety of 2D barcode symbologies have been developed and this trend is still progressing.

The recent trend in developing new 2D barcode systems can be broadly divided into two main streams: designing index 2D barcodes that can also be used as advertisements and developing database 2D barcodes maximised for data capacity within a given symbol space. The foci of the former approach are on designing an eye-catching 2D barcode and ensuring robust reading. In the latter approach, the use of colour elements has attracted the attention of researchers and developers since it can improve data capacity without increasing the number of cells within a given symbol space.

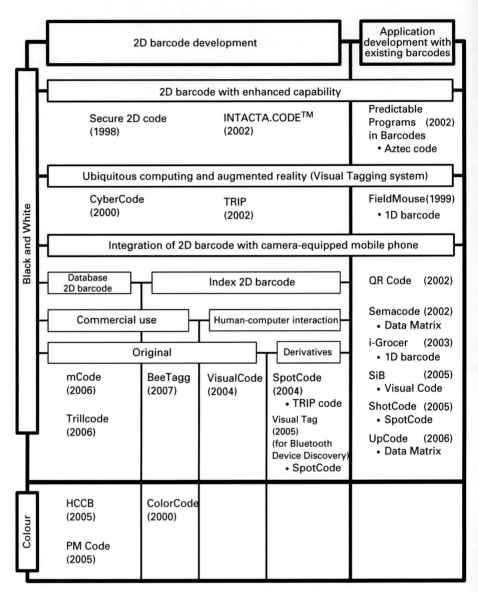

Fig. 4.19   The development of 2D barcode symbologies, and applications based on the existing symbologies. Each barcode is categorised in terms of the purpose of the development and the colour format; the year in which the barcode was introduced is given in parentheses. In the right-hand column the barcode from which an application was developed is given after a bullet point. For QR Code, 2002 was the year in which camera phones with a QR Code reader became available.

Starting with the taxonomic research reported in [15], in this chapter we have introduced the prolific research that has been carried out on barcode technology. The 2D barcode symbologies, and novel applications based on existing barcode symbologies, can be categorised in terms of the purpose of their development. Figure 4.19 illustrates the categories of 2D barcode symbologies and applications and their correspondences, along with the dates of developments. This figure also classifies them into two categories in terms of their colour format, i.e. black and white or colour.

In Chapter 5, technologies, techniques and schemes that can improve the robustness of 2D barcodes are described in detail.

5 Technologies for enhancing barcode robustness

One of the advantages of barcode technology is its fast, accurate and reliable operation. Most one-dimensional (1D) barcodes only use a checksum to ensure that the data are correctly decoded. The data of 1D barcodes are vertically redundant. This allows correct data retrieval even when the symbol has been partially damaged. When at least one horizontal path across the barcode is readable, a code with printing defects can be correctly read. Furthermore, human-readable characters are also printed below most 1D barcodes (see Figure 2.12). This allows users to input the data manually in the worst-case scenario. Hence, a checksum for error detection may be sufficient for a 1D barcode.

Moreover, two-dimensional (2D) barcodes do not have such vertical redundancy. Furthermore, they are not accompanied by human-readable characters, often because of the limitation in printing area for the data encoded in 2D barcodes. Printing the encoded data near the symbol may be possible when using index 2D barcodes with limited data capacity. However, considering the space efficiency advantage of 2D barcodes, it is reasonable not to print human-readable characters along with the 2D barcode symbols. In some cases, it might be impossible owing to space restrictions. This prevents users from correcting data manually if an error should occur. Hence, 2D barcodes need a means of not only detecting errors but also correcting them.

5.1 Error detection and correction codes

Among the many error detection and/or correction techniques, the one most commonly used for 2D barcodes is the Reed–Solomon code. Nearly all database 2D barcodes have adopted this technique. As mentioned above, the greater data capacity of a 2D barcode for a given space implies a corresponding reduction in the size of each cell, which makes accurate reading difficult. To enhance the reading robustness of such 2D barcodes, strong error detection and correction techniques are required. The Reed–Solomon code can satisfy such requirements.

5.1.1 Reed–Solomon codes

Reed–Solomon (RS) codes are block-based error correcting codes with a wide range of applications. These range from RS codes for CD-ROMs and DVDs, for data transmission technologies such as DSL and WiMAX and for broadcast systems such as those of digital video broadcasting (DVB) and the Advanced Television Systems Committee (ATSC) to the RS(255, 223) codes used by NASA in space communications. Reed–Solomon codes form a subset of the *Bose–Chaudhuri–Hochquenghem* (BCH) codes and are linear block codes with code symbols from the *Galois field* or a *finite field*. The code was invented by Irving S. Reed and Gustave Solomon in 1960. Reed–Solomon codes are robust and reliable error correcting codes and are particularly well suited to correcting burst errors where a series of bits in a codeword is received in error. However, the technology may not be ideal for systems that require fast operation since, owing to their complex implementation, Reed–Solomon codes are computationally more expensive than exclusive OR (XOR) based codes.

Basic principle of Reed–Solomon code

Like other error correcting codes, Reed–Solomon codes are based on the principle that redundancy is added to the original information so that errors due to data communication can be corrected [60]. A Reed–Solomon code is specified as RS(n, k) with r-bit symbols. This means that k data symbols each with r bits are encoded together with $n - k$ parity symbols, where each group of r bits is composed as an n-symbol codeword or block code. Reed–Solomon codes are capable of correcting up to t symbols that contain errors in a codeword, where

$$2t = n - k. \tag{5.1}$$

Figure 5.1 shows the structure of a codeword with Reed–Solomon codes in *systematic encoding*, where the data symbols always appear on the left unchanged and the parity symbols are appended to the right.

As an example, the parameter values for the most commonly used RS code (255, 223) with eight-bit symbols are

$$n = 255, \quad k = 223, \quad r = 8, \quad 2t = 32, \quad t = 16.$$

Thus, of the 255 symbols that compose a codeword, 223 are data symbols and the remaining 32 symbols are parity symbols added for error correction. With 32 parity symbols, up to 16 symbols that contain errors can be corrected.

Reed–Solomon codes are very effective in correcting bursts of bit errors as long as no more than t bytes are affected. Because Reed–Solomon codes correct byte errors, they

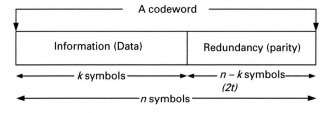

Fig. 5.1　　　　Structure of a codeword with Reed–Solomon codes in systematic encoding.

Table 5.1. (a) Addition and (b) multiplication in the finite field GF(2)

(a)

+	0	1
0	0	1
1	1	0

(b)

×	0	1
0	0	0
1	0	1

can potentially correct many bit errors. A symbol error occurs when one or more bits in a symbol are wrong. As mentioned above, RS(255, 223) can correct 16 symbol errors. Sixteen bit errors could occur, each in a separate symbol, sixteen complete symbol errors could occur. In the former case, the decoder needs to correct only 16 bit errors whereas 16×8 bit errors can be corrected in the latter scenario. This means that Reed–Solomon code is very good at correcting large clusters of errors [60].

Galois field arithmetic

Reed–Solomon codes are based on a specialist area of mathematics known as Galois fields (GFs) (also known as finite fields). For any prime number p, there exists a finite field denoted GF(p) that contains p elements.

The field GF(p) is called the *prime field* of order p, and is the field of residue classes modulo p; the p elements are $0, 1, \ldots, p-1$. A zero element, an inverse element and an identity element exist in the Galois field. Furthermore, the Galois field elements satisfy the commutative law, the associative law and the distributive law. Arithmetic operations performed in this field result in an element within the field. This allows the effective use of memory space when it is used for error correcting coding, since the result never exceeds the maximum element value.

The finite field GF(2) consists of the elements 0 and 1, which satisfy the addition and multiplication tables given in Table 5.1. Addition in GF(2) is the XOR operation and multiplication is the AND operation. Since the only invertible element is 1, division is the *identity function*. Table 5.1 shows that GF(2) is closed under addition, multiplication, subtraction and division.

It is possible to extend the field GF(p) to a field GF(p^r) of p^r elements, called an extension field of GF(p); r is a non-zero positive integer. The field GF(p^r) contains the elements of GF(p) as a subset. When $r > 1$, GF(p^r) can be represented as the field of equivalence classes of polynomials whose coefficients belong to GF(p). Any irreducible polynomial of degree r yields the same field up to an isomorphism. If an irreducible polynomial generates all the elements of GF(p^r) in this way, it is called a *primitive polynomial*. Primitive polynomials are used in the representation of elements of a finite field. If element α of GF(p^r) is a root of a primitive polynomial $f(x)$, the elements of GF(p^r) can be represented as successive powers of α:

$$GF(p^r) = 0, 1, \alpha, \alpha^2, \alpha^3, \ldots, \alpha^{p^r-2},$$

where $\alpha^{p^r-1} = \alpha^0$ (i.e. 1).

Table 5.2. Different representations of the elements of a finite field

Power	Polynomial	Vector	Integer
0	0	(000)	0
α^0	1	(001)	1
α^1	α	(010)	2
α^2	α^2	(100)	4
α^3	$\alpha + 1$	(011)	3
α^4	$\alpha(\alpha + 1) = \alpha^2 + \alpha$	(110)	6
α^5	$\alpha(\alpha^2 + \alpha) = \alpha^2 + \alpha + 1$	(111)	7
α^6	$\alpha(\alpha^2 + \alpha + 1) = \alpha^2 + 1$	(101)	5
α^7	$\alpha(\alpha^2 + 1) = 1$	(001)	1

Taking $GF(2^3)$ as an example, the modulus can be taken as either $x^3 + x^2 + 1$ or $x^3 + x + 1$. Using the modulus $x^3 + x + 1$, $f(x) = x^3 + x + 1$. Let α be a root of $f(x)$; thus

$$f(\alpha) = \alpha^3 + \alpha + 1 = 0. \tag{5.2}$$

Hence, the elements of $GF(2^3)$ can be represented as:

$$GF(2^3) = 0, 1, \alpha, \alpha^2, \alpha^3, \ldots, \alpha^6.$$

From Equation 5.2, $\alpha^3 = -\alpha - 1 = \alpha + 1$ (since $1 = -1$, in $GF(2)$). This allows the elements of $GF(2^3)$ to be represented as polynomials with degree less than 3. Table 5.2 presents the elements of the finite field $GF(2^3)$ in different representations: power (exponential), polynomial, vector and integer-corresponding-to-the-vector.

The set of polynomials in the second column is closed under addition and multiplication, modulo $\alpha^3 + \alpha + 1$. For any prime or prime power q and any positive integer r, a primitive irreducible polynomial of degree r over $GF(q)$ exists. The degree r determines the primitive polynomials. For Reed–Solomon codes, the finite field $GF(2^8)$ is often used.

Encoding Reed–Solomon codes

The maximum codeword length n for a Reed–Solomon code depends on the Galois field used for the arithmetic operations. When an extension field of degree r of $GF(2)$ or $GF(2^r)$ is used, every symbol becomes an element of $GF(2^r)$ with r bits, and the maximum n is then given by

$$n_{max} = 2^r - 1. \tag{5.3}$$

To generate a Reed–Solomon code, the Galois field and the error correction capability of the code t (i.e. $RS(n, k)$ with r bits per symbol) must be determined first.

Message polynomial

Once all the necessary parameters are determined, the message polynomial

$$m(x) = m_{k-1}x^{k-1} + \cdots + m_1 x + m_0$$

is encoded.

Generator polynomial

Then a Reed–Solomon codeword is generated using a special polynomial. All valid codewords are exactly divisible by the generator polynomial. The general form of the generator polynomial is

$$g(x) = \prod_{i=b}^{2t-1+b} (x - \alpha^i),$$

that is,

$$g(x) = (x - \alpha^0)(x - \alpha^1) \cdots (x - \alpha^{2t-1}),$$

where b is an integer; usually $b = 0$ or 1 [60].

Let $r = 8$, and let GF(2^8) be generated by a primitive element α with

$$f(\alpha) = \alpha^8 + \alpha^4 + \alpha^3 + \alpha^2 + 1.$$

Let $b = 0$, $t = 2$. Then there is an RS(255, 251) code with generator polynomial

$$g(x) = (x - 1)(x - \alpha)(x - \alpha^2)(x - \alpha^3)$$
$$= x^4 + \alpha^{75}x^3 + \alpha^{249}x^2 + \alpha^{78}x + \alpha^6.$$

Codeword

The codeword $c(x)$ is constructed using

$$c(x) = x^{n-k}m(x) + p(x),$$

where

$$p(x) = x^{n-k}m(x) \quad \mathrm{mod}\ g(x).$$

Decoding Reed–Solomon codes

In decoding a Reed–Solomon code, we must first

(i) determine whether any errors exist, using syndrome computation.

If any errors are found then we must

(ii) detect the number of symbols that contain errors;
(iii) detect the location of the symbols that contain errors;
(iv) evaluate the error values; and
(v) correct the errors.

Syndrome computation
The received codeword $r(x)$ is given by

$$r(x) = c(x) + e(x), \qquad (5.4)$$

where $c(x)$ is the original codeword and $e(x)$ is the error. The *syndrome* can be computed using this relationship. Furthermore, $c(x)$ is divisible by $g(x)$; thus, the syndrome is given by

$$S(x) = r(x) \bmod g(x)$$
$$= (c(x) + e(x)) \bmod g(x)$$
$$= e(x) \bmod g(x).$$

A Reed–Solomon codeword has $2t$ syndromes, which depend only on the error [61]. The syndrome of the received polynomial is:

$$S_i = r(\alpha^{i-1-b}) \qquad (i = 1, 2, \ldots, 2t). \qquad (5.5)$$

Error detection can be performed by examining the syndrome polynomial. If $S_i = 0$, then no error is detected. If there exists $\exists S_i \neq 0$ for $1, 2, \ldots, 2t$ then this indicates that $r(x) \neq c(x)$; thus that $r(x)$ contains errors.

Finding the number of errors
When errors are found, the number of errors must be determined first. The *Peterson algorithm* can be used to determine the number of errors in $r(x)$ when that number is small, and it is described below. Suppose that there are k errors; then the following matrix can be generated using the syndromes given in Equation 5.5:

$$\begin{vmatrix} S_k & S_{k-1} & \cdots & S_1 \\ S_{k+1} & S_k & \cdots & S_2 \\ \vdots & \ddots & \vdots & \vdots \\ S_{2k-1} & S_{2k-2} & \cdots & S_k \end{vmatrix} \neq 0. \qquad (5.6)$$

When the number of errors is not equal to k, Equation 5.6 returns 0.

Assume that $k = t$ and generate the matrix in the same way. If the determinant is 0 then try $k = t - 1$ and repeat the operation. When Equation 5.6 returns a value other than 0, this indicates that the assumed number of errors is equal to the actual number of errors [61].

Detecting the error locations
Several fast algorithms such as the Peterson algorithm, the Euclidean algorithm or the Berlekamp–Massey algorithm can be used to determine an error-locator polynomial. In practice, the Euclidean algorithm has been widely used owing to its easy implementation. However, it has been demonstrated that the Berlekamp–Massey algorithm allows fast and extremely efficient decoding of dozens of symbol errors [60]. Despite its simple

operation, the computational complexity of the Peterson algorithm is very high when a large number of errors is to be corrected.

Using the Peterson algorithm, the locations of the k errors can be determined by generating the following error-locator polynomial [61]:

$$\sigma(x) = 1 + \sigma_1 x + \sigma_2 x^2 + \cdots + \sigma_k x^k,$$

where $\sigma_1, \sigma_2, \ldots, \sigma_k$ satisfy the matrix equation

$$\begin{bmatrix} S_k & S_{k-1} & \cdots & S_1 \\ S_{k+1} & S_k & \cdots & S_2 \\ \vdots & \ddots & \vdots & \vdots \\ S_{2k-1} & S_{2k-2} & \cdots & S_k \end{bmatrix} \begin{bmatrix} \sigma_1 \\ \sigma_2 \\ \vdots \\ \sigma_k \end{bmatrix} = \begin{bmatrix} S_{k+1} \\ S_{k+2} \\ \vdots \\ S_{2k} \end{bmatrix}. \tag{5.7}$$

When $\alpha^0, \alpha^1, \ldots, \alpha^{n-2}$ are substituted into $\sigma(x)$, there exist k locations where $\sigma(x) = 0$. The error positions are yielded as the inverse roots of $\sigma(x)$. For example, if $\sigma(\alpha^j) = 0$ then this indicates that the coefficient of x^{255-j} in $r(x)$ contains errors.

Evaluating the error values

To compute the error values using the Peterson algorithm, the syndromes in Equation 5.5 can be used. From $r(x) = c(x) + e(x)$ (see Equation 5.4), the syndrome can be written as

$$S_i = c(\alpha^{i-1+b}) + e(\alpha^{i-1+b}) \qquad (i = 1, 2, \ldots, 2t). \tag{5.8}$$

If $j_1, \ldots, j_k$ correspond to the errors, for example, we can obtain expressions for the error values $e_1, \ldots, e_k$ using Equation 5.8:

$$\begin{cases} S_1 = r(\alpha^b) = \alpha^{j_1 b} e_1 + \alpha^{j_2 b} e_2 + \cdots + \alpha^{j_k b} e_k, \\ S_2 = r(\alpha^{b+1}) = \alpha^{j_1(b+1)} e_1 + \alpha^{j_2(b+1)} e_2 + \cdots + \alpha^{j_k(b+1)} e_k, \\ \vdots \\ S_k = r(\alpha^{b+k-1}) = \alpha^{j_1(b+k-1)} e_1 + \alpha^{j_2(b+k-1)} e_2 + \cdots + \alpha^{j_k(b+k-1)} e_k, \end{cases}$$

Correcting the errors

Using the error values $e_1, \ldots, e_k$, the errors are corrected using

$$c(x) = r(x) + x^{j_1} e_1 + x^{j_2} e_2 + \cdots + x^{j_k} e_k. \tag{5.9}$$

5.1.2   Other error detecting and correcting techniques

Unlike database 2D barcodes some 2D barcodes with low data capacity, such as index 2D barcodes, use other error detection and/or correction codes. Although 2D barcodes do not have the vertical redundancy found in 1D barcodes, the cell sizes of these codes are usually sufficiently large to be accurately located and read. Also, the limitation in data capacity limits the choices of possible error detection and/or correction codes. Hence, these 2D barcodes use error correcting codes that require less capacity, such as Hamming codes and parity-check codes.

Hamming codes

A Hamming code is a linear error correcting code named after its inventor, Richard Hamming. Hamming codes can detect and correct single-bit errors and can detect, but not correct, up to two simultaneous bit errors. Unlike Reed–Solomon codes, a Hamming code is not capable of correcting burst errors. However, owing to their simple and fast encoding and decoding, Hamming codes, or binary BCH codes with a minimum distance 3, have been very popular in computer networks and in memory devices [60] where the probability of error is low. Visual Code uses the Hamming code.

The length n of a Hamming code and the number k of its data bits are given by

$$n = 2^m - 1$$

and

$$k = n - m,$$

where m denotes the number of parity-check bits. For example, with $m = 3$ we obtain the Hamming (7, 4, 3) code.

Hamming codes use two matrices for their operation, a parity-check matrix and a generator matrix. The parity-check matrix H is an $m \times n$ matrix that contains non-zero distinct columns. When the columns of H are binary vectors of length m, there are up to $2^m - 1$ possible non-zero distinct columns, which defines the length n of a Hamming code. As an example, the parity-check matrix H for the Hamming (7, 4, 3) code is

$$H = \begin{bmatrix} 1 & 1 & 0 & 1 & 1 & 0 & 0 \\ 1 & 0 & 1 & 1 & 0 & 1 & 0 \\ 0 & 1 & 1 & 1 & 0 & 0 & 1 \end{bmatrix}.$$

The order of the columns is arbitrary. When we set

$$H = [A \ \ I_{n-k}],$$

the matrix H is in its *systematic* form and contains an arbitrary matrix A and the $(n - k) \times (n - k)$ identity matrix I_{n-k}.

The generator matrix G must satisfy

$$HG^T = GH^T = 0, \tag{5.10}$$

where $G \neq 0$. The first four columns of the systematic parity-check matrix H are all the non-zero n-*tuples* and the last three are the $(n - k) \times (n - k)$ identity matrix I_{n-k}.

Hence, G can be obtained from H by taking the transpose of the first four columns of H with the $k \times k$ identity matrix I_k,

$$G = [I_k \ A^T].$$

The resultant generator matrix G is given as

$$G = \begin{bmatrix} 1 & 0 & 0 & 0 & 1 & 1 & 0 \\ 0 & 1 & 0 & 0 & 1 & 0 & 1 \\ 0 & 0 & 1 & 0 & 0 & 1 & 1 \\ 0 & 0 & 0 & 1 & 1 & 1 & 1 \end{bmatrix}.$$

Encoding

In this code there are 2^k possible codewords. When $k = 4$, as in the above matrix the number of codewords is $2^4 = 16$. The codeword vector $\vec{c}$ is constructed using the generator matrix G and the message bit vector $\vec{m} = [m_1 \cdots m_k]$ as

$$\vec{c} = \vec{m}G.$$

For example, when encoding the message bit vector $[1\,0\,1\,1]$, the codeword is computed as

$$\begin{bmatrix} 1 & 0 & 1 & 1 \end{bmatrix} \begin{bmatrix} 1 & 0 & 0 & 0 & 1 & 1 & 0 \\ 0 & 1 & 0 & 0 & 1 & 0 & 1 \\ 0 & 0 & 1 & 0 & 0 & 1 & 1 \\ 0 & 0 & 0 & 1 & 1 & 1 & 1 \end{bmatrix} = \begin{bmatrix} 1 & 0 & 1 & 1 & 0 & 1 & 0 \end{bmatrix}.$$

Hence, the generated codeword $\vec{c}$ is $[1\,0\,1\,1\,0\,1\,0]$.

Decoding

Hamming codes contain the property of their parity-check matrix H of having distinct columns as the binary representation of integers. If a single error occurs in position i, $1 \leq i \leq n$, then the syndrome of the received vector equals the column of H in the position in which the error occurred. Let $\vec{e}$ denote the error vector that has been added in the transmission of a codeword and assume that its components are all equal to zero except for the ith component, where $e_i = 1$. Then the syndrome of the received codeword $\vec{r}$ is given by

$$\vec{S} = \vec{r}H^T = \vec{e}H^T = \vec{h}_i,$$

where $\vec{h}_i$ denotes the ith column of H and $1 \leq i \leq n$. This is the key operation in decoding.

In detail, when no error has occurred the received codeword is

$$\vec{r} = \vec{m}G,$$

where $\vec{m}$ denotes the original message bits. From Equation 5.10,

$$\vec{r}H^T = \vec{m}(GH^T) = 0.$$

This indicates that no error is found when the product of the received word and the parity-check matrix H is zero, and vice versa. Suppose that $\vec{r}$ contains a one-bit error; then it can be expressed as

$$\vec{r} = \vec{m}G + e_i,$$

where e_i, the error at position i, equals 1 and all other components equal zero. The product of the received codeword and the transpose of the parity-check matrix is

$$\vec{r}H^T = (\vec{m}G + e_i)H^T = \vec{m}(GH^T) + e_iH^T = e_iH^T = \vec{h}_i.$$

For example, when the received codeword $\vec{r} = [1\ 1\ 1\ 1\ 0\ 1\ 0]$ contains an error, the position of the error can be detected by

$$\begin{bmatrix} 1 & 1 & 1 & 1 & 0 & 1 & 0 \end{bmatrix} \begin{bmatrix} 1 & 1 & 0 & 1 & 1 & 0 & 0 \\ 1 & 0 & 1 & 1 & 0 & 1 & 0 \\ 0 & 1 & 1 & 1 & 0 & 0 & 1 \end{bmatrix}^T = \begin{bmatrix} 1 & 0 & 1 \end{bmatrix}.$$

The transpose of the outcome [1 0 1] equals the value of the second column of H, indicating that the second column of the received word has an error. By correcting it, the original codeword $\vec{c} = [1\ 0\ 1\ 1\ 0\ 1\ 0]$ can be obtained.

Parity

Parity code is a special case of Hamming code. It provides a self-checking feature in barcodes and other data transmission techniques. A parity bit can detect an odd number of errors. However, unlike the Reed–Solomon and Hamming codes, a simple parity code is not capable of correcting errors; hence, it is only useful when more than two errors rarely occur. As an example, the parity-check method was adopted for the index 2D barcode ColorCode, to ensure data integrity.

Parity code adds a single *check bit* (also called a redundancy bit) that indicates whether the number of '1' bits in the preceding data was even or odd. If an odd number of bits is changed during data transmission, the value of the parity bit will become invalid, indicating that an error has occurred. There are two different parity protocols: even parity and odd parity. Adding a single bit to the binary array, the even-parity protocol makes the sum of all the bits always even, whereas the odd-parity protocol makes it odd. The even-parity protocol determines that no error has occurred when the number of 1s in the received data is even, and vice versa.

Parity checking is not robust. If the number of bits changed is even, for example if a 1 has been converted to 0 and a 0 has been converted to 1 in a given set of bits, no change will be detected and the received data will be considered to be valid. It is also possible that the changed bit may have been the parity bit itself. Furthermore, parity does not indicate which bit contains the error, even when it is detected.

Data	7 bits of data							Pb
D1	0	1	0	0	0	0	1	1
D2	1	0	0	0	0	0	1	1
D3	0	1	0	0	1	0	1	0
D4	1	1	0	0	0	0	1	0
D5	1	1	1	1	0	0	1	0
D6	0	0	1	0	0	0	1	1
D7	1	0	1	0	0	0	1	0
Pb	0	0	1	1	1	0	1	1

LRC using odd parity

TRC using even parity

Fig. 5.2   Example of parity checking; D1–D7 are seven different sets of character data and Pb is the parity bit.

In practice, a combination of a *longitudinal redundancy check* (LRC) (also known as a horizontal redundancy check) and a *transverse redundancy check* (TRC) (or vertical redundancy check) is used to improve the parity-checking performance (see Figure 5.2). For an LRC the data must be divided into transmission blocks, to which the parity bit is added. However, the TRC is a way of error checking by adding a parity bit to each byte of data to be transmitted.

5.2   Effectiveness of data arrangement

Error detecting and/or correcting codes directly improve the robustness of barcode reading. However, there is a trade-off between the capability of the error correcting codes and the data capacity of a 2D barcode. An increase in the number of error correcting codes within a given fixed space means a decrease in the space available for encoding data. Furthermore, powerful error correcting codes often slow down the barcode decoding process.

There are techniques that are computationally inexpensive yet help to enhance the robustness of barcode reading. Examples are data arrangement techniques such as interleaving and masking. On their own they do not directly improve robustness. However, in combination with other techniques they work very efficiently. For example, with interleaving, Reed–Solomon codes can improve performance significantly. In Section 5.2.1 we will explain how.

5.2.1   Interleaving

Interleaving is often used in data communication to protect data integrity against long contiguous burst errors. By rearranging the array of the original data in a non-contiguous way, interleaving allows consecutive errors to become random errors, which can be corrected by the error correcting technique employed.

To improve the robustness of barcode reading, an error correcting code such as the Reed–Solomon code is normally used. There is a trade-off between the robustness of a 2D barcode symbol and its data capacity. Although encoding more error correcting codes improves the tolerance of errors, it reduces the space for encoding information data. Hence, the error correction level is often determined according to the environment in which a 2D barcode symbol is to be used. When a barcode is less likely to be damaged, a lower level of error correction capability may be sufficient.

The amount of error correcting codes used determines the number of correctable errors. When the number of errors in a single codeword exceeds the number of correctable errors, the original data cannot be successfully restored, resulting in decoding failure. This is more likely when errors come in long bursts. Interleaving prevents it from happening by de-interleaving the received data, which converts the long burst errors into random errors. Figure 5.3 demonstrates the difference in performance of data restoration with and without interleaving. As a result, the number of errors in a codeword becomes within the

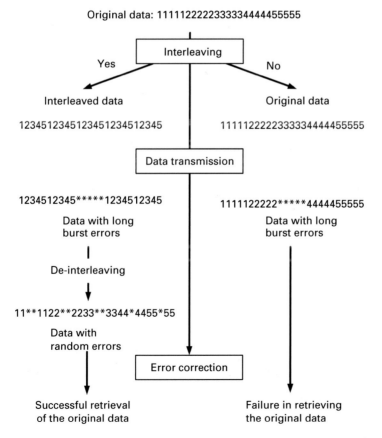

Fig. 5.3 Effect of interleaving.

capability of the error correcting technique employed, resulting in successful retrieval of the original data. Interleaving does not involve complex computations. Furthermore, it helps to improve the robustness of barcode reading without increasing the power of error correction. Hence, it is a cost-effective technique for improving robust reading.

5.2.2   Masking

Masking is a technique used in QR Code to enhance the robustness of barcode symbol reading by even allocation of the black and white cells.

When a chunk of black cells with only few white cells is concentrated in a particular portion of a QR Code symbol, this may cause scanning devices to miss some white cells or vice versa; the result is a decoding failure. A data area having a black and white pattern similar to that used in the QR Code finder pattern (i.e. $1:1:3:1:1$) is also undesirable since it could confuse the reading software and slow down its performance. Masking can be used to minimise such a concentration of one colour and to avoid finder-pattern-like cell formations within the data area, so that the performance of the QR Code decoding can be improved.

As illustrated in Figure 5.4, there are eight different pre-defined mask patterns. Every mask pattern is applied to the module pattern (i.e. the black and white cells before they are masked) of the data area of a QR Code symbol, and then the suitability of each masking of the symbol is evaluated to determine the best mask pattern for that particular barcode symbol.

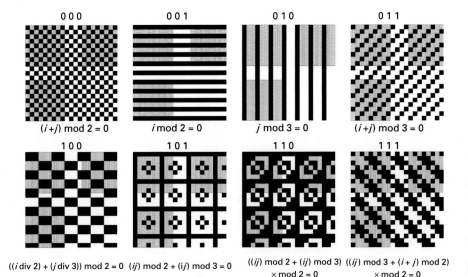

Function module Note that masking is not applied to the function modules.

Fig. 5.4   The eight QR Code mask patterns; showing the pattern size used for version 1 of QR code; (i, j) are the cell (module) coordinates; starting with $(0, 0)$ for the top left cell (adapted from [27]).

Table 5.3. Masking evaluation criteria and deducted points

Undesirable cell pattern	Evaluation criteria	Points deducted[a]
Sequence of one-colour cells either in a row or a column	Total number of such cells $= 5 + i$	$P_1 + i$
Block of one-colour cells	Block size $= 2 \times 2$	P_2
Finder-pattern-like cell pattern (i.e. B : W : B : W : B = 1 : 1 : 3 : 1 : 1)	Four consecutive white cells appeared before and after finder-pattern-like cells	P_3
Percentage of black cells in the whole symbol	$(50 \pm 5q)\% - (50 \pm 5(q + 1))\%$	$P_4 q$

[a] $P_1 = 3$, $P_2 = 3$, $P_3 = 40$, $P_4 = 10$.

Masking evaluation takes place as follows.

(i) The candidate symbol is masked by applying bitwise-XOR operations to the cell pattern (or module pattern), using each pre-defined mask pattern in turn. By this operation, a black cell of the candidate symbol corresponding to a black cell of the mask pattern will be converted to a white cell. The reverse applies to the white cells of the candidate symbol. That is, a white cell that corresponds to a black cell of the mask pattern will be converted to a black cell.

(ii) Each masked symbol is evaluated by reducing its suitability score when any undesirable cell pattern is found.

(iii) The mask pattern that retains the highest suitability score is selected.

The evaluation criteria and an example of the QR Code masking operation are presented in Table 5.3 and Figure 5.5, respectively.

Four different undesirable cell patterns are defined. The variables $P_1 - P_4$ in Table 5.3 indicate the points to be deducted for respective undesirable patterns. The variable i denotes the number of cells by which a sequence of one colour exceeds five cells. For example, if a sequence of seven black cells is found, the sum of P_1 (i.e. 3) and i (i.e. $7 - 5 = 2$), namely, $3 + 2 = 5$, should be deducted. The variable q is obtained by calculating the magnitude of the deviation, i.e. the difference between the percentage of black cells of the candidate symbol and the base percentage of 50%. Ten points are deducted for every 5% difference from the base percentage. For example, when the black cells account for 45% of the candidate symbol, the difference from the base percentage will be 5%, resulting in 10 points reduction.

5.3 Symbol structuring

The techniques introduced in the previous sections are an attempt to improve barcode robustness by data manipulation. These techniques are only applied to the data. Making other improvements to the structure of barcode symbols can also help enhance barcode robustness. For example, it is necessary for most barcodes to have a quiet zone to assist

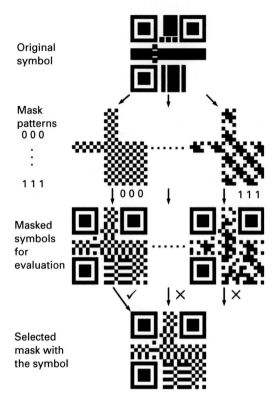

Original
symbol

Mask
patterns
0 0 0

.
.
.

1 1 1

↓ 0 0 0   ↓   ↓ 1 1 1

Masked
symbols
for
evaluation

✓   ↓ ✕   ↓ ✕

Selected
mask with
the symbol

Fig. 5.5   QR Code masking operation (adapted from [27]).

the detection and location of a barcode symbol among its surrounding objects. Some objects may have a similar shape to a barcode and so may confuse the scanning and decoding devices. A quiet zone or clear space is used to separate a barcode symbol from such background. It also helps the decoding software decide when to start and stop its operation. The quiet zone in 2D barcodes surrounds the symbol, whereas 1D barcodes usually have a leading and a trailing quiet zone before the start character and after the stop character, respectively. This is due to the difference in reading 1D barcodes and 2D barcodes. Whereas the former are read across the black and white bars, nearly all matrix 2D barcodes are read omnidirectionally, which is why a quiet zone surrounding the latter type of barcode is needed.

As its name depicts, a critical component in the location of a barcode symbol is its finder pattern (also known as a position-detection pattern). In fact, it is one of the keys for successful reading of a barcode symbol.

5.3.1   Finder pattern

Unlike human eyes, scanning devices or decoding software cannot distinguish a candidate barcode symbol from its background. The main task of a finder pattern is to inform the scanning and decoding device where the candidate symbol is. To design the best finder pattern for a particular barcode is very important for symbol detection and robust barcode reading. Finder patterns are commonly used for:

(i) detecting and locating the barcode symbols;
(ii) correcting the orientation, distortion and tilt of the symbols;
(iii) measuring the size of the entire symbol, which enables measurement of the cell size; and/or
(iv) computing of the number of cells.

The role of the finder pattern is crucial for accurate barcode decoding. The best finder pattern for a particular barcode symbology depends on the purpose of its use or application, the tasks of the finder pattern, the symbol design and so forth. For example, a central bull's eye was selected as the finder pattern for ShotCode and MaxiCode since the initial purpose of these 2D barcodes was to handle moving targets among other objects. It was found that a circular shape worked best. The work in [20] stated:

[A] Circular shape works better for the identification and 3D localisation of the object than the other shapes (e.g., square) using a visual sensor. In man-made environments, circles are less common shapes than right angles, squares and rectangles. TRIP is intended for the localisation of tagged objects in these highly cluttered environments, where the central bull's eye of a TRIP tag represents a very distinctive pattern. The detection of squares in those cluttered environments supposes an expensive computational task given the many straight edge combinations possible within an image. An even more unusual shape could have been chosen, for example, a star. In this case, however, there is the problem of representing such a shape in a compact way and to encode useful information (e.g., ID) around or within it. A circle, and its projection in an image as an ellipse, can be represented in a compact and elegant form by a symmetric (3×3) matrix. Moreover, ellipses in an image are salient features and can be detected robustly and rapidly.

In contrast, mCode uses an unobtrusive finder pattern made up of a combination of dot-like cells or blobs. Such a finder pattern is not only space-effective but also suitable for advertising applications, allowing mCode to be integrated with commercial or artistic images such as a company logo without its appearance as a whole being spoiled. mCode was invented with the aim of developing a database barcode suitable for advertisements. Hence, the blob finder pattern is ideal for mCode.

Similarly, the aim of the QR Code development was to provide a database 2D barcode that can be read ultra-fast, handling not only stationary objects but also moving targets on a conveyor in industrial use. The resultant finder pattern is made up of distinctive squares at the three corners of the symbol, each being composed of three different-sized black and white squares. The unique ratio of black and white (i.e. $1:1:3:1:1$) on a line passing through the centre of the finder pattern at any angle enables fast detection of the symbol.

However, there is one thing that should be kept in mind when designing or selecting a finder pattern for 2D barcodes, i.e. the impact on the reading robustness when the finder pattern gets damaged. With error correcting codes such as the Reed–Solomon code, the damaged data can be corrected. However, the finder pattern cannot be restored once it is damaged [13], which could slow down the decoding operation or, in the worst case,

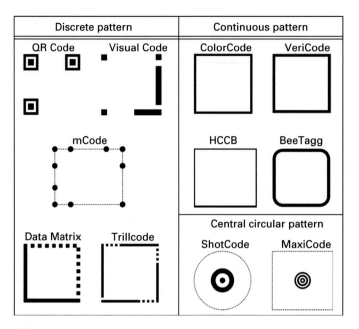

Fig. 5.6 Finder pattern examples.

may cause a symbol detection failure. Hence the finder pattern should be designed so that it will still function properly even when it is partially damaged.

A variety of finder patterns have been designed and they are categorised into three groups: *discrete*, *continuous* and *central circular* patterns. Figure 5.6 presents examples of such finder patterns.

Square finder patterns with either a solid border (e.g. ColorCode and VeriCode) or a combination of solid and broken borders (e.g. Data Matrix and Trillcode) are most commonly used. When small dots are used to compose a finder pattern, losing even a single dot may confuse the decoding program. When increasing the data density of a barcode for a given space, so that the barcode cell size decreases, the size of the finder pattern also decreases. However, a very small finder pattern may be missed by the reader even when it is not damaged. Designing a finder pattern similar to the one used by QR Code, which is rather large, may be a possible solution. However, increasing the size of the finder pattern would decrease the space available for encoding data, resulting in less data capacity for a given symbol size. To ensure the robustness and reliability of the barcode, it may be reasonable to design or select a solid continuous border finder pattern unless this conflicts with the purpose of the particular barcode.

5.3.2 Use of supplementary means

In general, the data capacity of index 2D barcodes is low and fixed, which allows scanning devices to measure the size of cells accurately and compute the centroid of each cell, resulting in accurate reading of the barcode. Database 2D barcodes, however, are scalable in data capacity. The amount of data can be increased according to the requirement of the application until it reaches the encoding limit of a particular barcode. The more data are encoded, the larger the symbol size will need to be, to increase the number of data cells.

As previously stated, the finder pattern is used not only for detecting a barcode symbol but also for measuring the sizes of the symbol and its cells. Even when the measurement is done properly, there is still the possibility of a slight difference between the computed cell size and its actual size. The difference might be negligible for a small symbol with a small amount of data, causing little or no negative effect when decoding the symbol. However, when the differences accumulate across a larger symbol, the decoding software may end up reading a value at the wrong location (e.g. at the border of two cells instead of a cell centre).

In fact, it was observed that the first-read rate of Data Matrix ECC 000–140 dropped dramatically when the data capacity, and consequently, the symbol size increased [10]. To remedy this, a block structure was added to the newer version of Data Matrix code, namely, ECC 200. In this version, a symbol is divided into blocks once its size exceeds 24 × 24 modules (i.e. cells) and solid borders are inserted between the blocks, separating them. The borders are to help correct symbol distortion as well as to fine-tune the computed positions of the cells. An example is presented in Figure 3.14.

Quick Response (QR) Code uses alignment patterns and timing patterns to solve problems such as symbol distortion and cumulative error. The alignment pattern consists of three concentric squares, each with 5 × 5 black cells, 3 × 3 white cells and one black cell. The number of alignment pattern blocks and their coordinates are pre-defined for every symbol version. The alignment pattern blocks are arranged symmetrically along the diagonal line between the top left and the bottom right of the symbol. A black cell within each alignment pattern block enables fast calculation of its centre coordinate, which is used to correct for local distortion of the symbol.

In QR Code there are both horizontal and vertical timing patterns. Each timing pattern is made up of alternating single black and white cells, starting and ending with a black cell. The sixth row between the separators of the top left and top right finder pattern blocks is occupied by the horizontal timing pattern and, similarly, the sixth column is occupied by the vertical timing pattern [27]. The data capacity of the symbol and the version information can be computed from the size of the timing patterns. These patterns are also used to obtain the coordinates of the centroids of the corresponding data cells. Any changes in cell pitch are corrected using the coordinates indicated by the timing patterns. Figure 3.4 shows the alignment pattern and the timing pattern.

High Capacity Color Barcode (HCCB) also uses single-pixel white borders (space separators) between the cells, but for a different purpose. For a colour 2D barcode such as HCCB, it is crucial to retrieve colours as close as possible to the original colour values. The white separator (or white spacing) is used to reduce the effects of anti-aliasing, and this results in more accurate colour sampling.

When developing either a monochrome or colour 2D barcode, various strategies should be used to maximise its robustness. Owing to their scalability in data capacity, database 2D barcodes can require supplementary means to ensure accurate data retrieval when compared with index 2D barcodes.

5.3.3   Structured append

Structured append allows multiple 2D barcode symbols to be logically linked together in order to encode a large amount of data. This means that a large symbol can be divided into multiple smaller symbols when it cannot be placed in a given area owing to space limitation. This feature can be used to manipulate the size of data cells when it comes to 2D barcodes for mobile devices. The cell size is one of the keys for robust barcode reading. Barcode readers either fail to retrieve the original data encoded in a barcode or misread it when the size of the cells is smaller than the size detectable by mobile devices. By dividing a symbol into multiple symbols, the structured append can increase the cell size in each structured append symbol.

In order to reconstruct the original data regardless of the reading order of the multiple symbols, additional information is encoded into each symbol. Taking QR Code as an example (see Figure 5.7), whether a symbol is part of the structured append feature is indicated by the *structured append mode indicator* in the header block included in the first

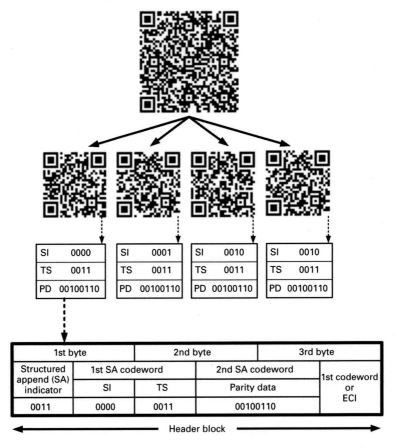

Fig. 5.7   QR Code structured append feature. SI, sequence indicator; TS, total symbols; PD, parity data; ECI, extended channel interpretation.

three data cells, each with 1 byte [27]. If the symbol is part of a structured append, '0011' is encoded as the first four bits of the first byte. It is followed by two *structured append code-words*, comprising the last four bits of the first byte, the second byte and the first four bits of the third byte. The first four bits of the structured append codeword form the structured append sequence indicator m and the second four bits give the total number of structured append symbols n. The binary values of $m - 1$ and $n - 1$ are encoded, respectively.

The second structured append codeword contains the structured append parity data. This value is calculated from all the data that are broken down into structured append symbols. Consequently, each structured append symbol has identical parity data, which ensures that all the structured append symbols are derived from an identical source. The parity data value is obtained by XORing byte by byte the ASCII/JIS values of all the original input data before dividing the data into multiple structured append symbols.

In Figure 5.7, a large QR Code symbol is divided into four smaller structured append symbols; the number of smaller symbols is indicated by the value in the TS box in Figure 5.7. By reading the data in the order of the values in the sequence indicator (SI), the original data can be reconstructed.

Data Matrix also has this feature. A single data symbol of Data Matrix, like QR Code, can be divided into 16 smaller symbols.

5.3.4 Use of reference cells

Some barcodes include reference cells to ensure accurate data retrieval. Format information such as barcode version, data type and masking pattern are often encoded in the reference cells. For colour 2D barcodes to be robust and reliable, it is vital to use reference cells that can provide a key to the colours used in encoding the data.

As explained earlier, a colour image could be reproduced differently from its original colour, depending on the image capturing and processing device (e.g. scanner, digital camera or webcam), the displaying device, such as a monitor, the printing device and the quality of the paper on which the image is printed. The lighting conditions can also have a great impact on the colour values of the captured or reconstructed images. When illumination is not applied evenly to an image, a particular colour value in the image may be reconstructed as different colour values, owing to the lighting effects. Colour cells may also fade and degrade over time, thus changing the colour values.

Hence, the black and white barcode format was preferred by most developers when designing their original barcodes. For example, in [17] it was mentioned that the changes in colour values caused by different lighting conditions constituted the reason why a black and white format was chosen for Visual Code. In [20] reference was made not only to the lighting issue but also to the differences in colour sensitivity of various CCD devices as being problematic for the robust decoding of a colour barcode. The black and white barcode format is often used because these colours are the furthest distance apart within a given colour space. The task of a decoding device is to discriminate two such colours, resulting in robust and reliable decoding even when the low-resolution camera of a mobile phone is used. However, the advantage of using black and white barcodes comes at the cost of reduced data capacity [42].

In ColorCode this problem was overcome by the use of reference cells that provide a standard colour for distinguishing each reproduced colour cell. The value of each colour cell in the data area is determined relative to the values of the corresponding standard colour in the reference cell. Since the relative difference in the colour value of a data cell from that of the standard colour is consistent regardless of the devices used for image capturing, displaying or printing, the data can be retrieved correctly. This idea and its successful implementation have encouraged the development of 2D barcodes that use a colour format. For example, the newer colour 2D barcode HCCB has also added a colour reference area or 'reference palette' that allows colour calibration between the colour cells in the data area and in the reference palette. Paper Memory (PM) Code also includes a colour reference area within the symbol.

This demonstrates that the use of reference colours for colour calibration is essential for colour barcodes and should be adopted as a first step to improve their robustness. Careful consideration may also be required in determining the size, shape and location of the colour reference cells.

There are many useful techniques to improve the robustness of 2D barcodes. A good understanding of these techniques is needed to develop a robust and usable 2D barcode.

5.4   Summary

One of the advantages of barcode technology is its fast, accurate and reliable operation. The data of 1D barcodes are vertically redundant, which allows correct data retrieval even when the symbol has been partially damaged. Furthermore, human-readable characters are also printed below the 1D barcodes, allowing users to input the data manually in the worst-case scenario. Hence, a checksum for error detection may be sufficient for a 1D barcode to operate reliably.

However, 2D barcodes have neither such redundancy nor means to correct data manually when error occurs. This means that 2D barcodes require built-in error detection and self-correction. To improve their usability and reliability, most database 2D barcodes use a robust and reliable error correcting technique such as the Reed–Solomon code. The limitation in data capacity of index 2D barcodes limits the choice of possible error detection and correction codes. Hence, a Hamming code and a parity-check code are often adopted for these 2D barcodes.

There are other ways to improve the robustness of 2D barcodes, for example data arrangement techniques such as interleaving and masking. With interleaving, the performance of an error correcting code (e.g. the Reed–Solomon code) can be significantly improved. Masking helps to enhance the robustness of barcode symbol reading by allocating black and white cells more evenly. Unlike error detecting and correcting codes, these techniques do not actively improve a barcode's robustness. However, they are efficient and cost-effective supporting players in enhancing the overall robustness of barcode reading.

Thus, one way of improving the robustness of 2D barcodes is to manipulate the data using techniques such as error correcting codes and data arrangement. Another way is designing a more functional 2D barcode and symbol structure. In particular, designing a robust and functional finder pattern is vital for the successful reading of 2D barcode symbols since the reading software cannot perform its subsequent tasks without first locating the barcode symbol. The use of additional means such as alignment patterns and timing patterns can help correct for symbol distortion. This allows the reading software to sample the values of data cells at the correct locations, resulting in more accurate data retrieval.

Structured append allows a large barcode symbol to be divided into several smaller symbols or several small symbols to be combined into one large symbol. This feature can be used to manipulate the size of data cells when it comes to 2D barcodes for mobile devices. When the cell size is smaller than the size detectable by mobile devices, the reading software either fails to retrieve the original data encoded in a barcode or misreads it; thus, an appropriate cell size is another key to robust barcode reading. The structured append feature is an example of an effective symbol restructuring scheme.

A colour image could be reproduced differently from its original colour, depending on the image capturing and processing devices, the displaying devices, the printing devices and the quality of paper on which the image is printed. Lighting also significantly affects the values of colours. Thus, for a colour 2D barcode, using colour reference cells is essential for successful barcode reading. When designing a colour 2D barcode, optimising the size, shape and location of the colour reference cells is important for enhancing the robustness of the barcode reading.

Many technologies and techniques can be used to improve the reading robustness of 2D barcodes. A good understanding of them will help in developing a robust and usable 2D barcode. In Chapter 6 we introduce a novel colour 2D barcode and its encoding and decoding algorithms. In that chapter we also discuss different methods for modelling colour, possible ways of colour selection, the different colour channels involved in a colour barcode system and ways of implementing a robust and reliable colour 2D barcode system through the development of our novel 2D barcode.

6 A prototype colour 2D barcode development

6.1 Aims of the prototype colour 2D barcode development

As mentioned in Chapter 2, most 2D barcodes were developed to meet special requirements. For example, QR Code was invented in response to the demand for a barcode with greater data capacity than the traditional linear barcodes so that it could collect information about products without accessing a backend database and hence improve the productivity in assembly lines. Visual Code was developed to implement applications for human–computer interaction.

To design a barcode, first of all the purpose of barcode development and the target applications must be determined. The ultimate goal of our barcode development was to implement a 2D barcode system that can provide applications and services ubiquitously, regardless of time, place and occasion. To implement such a system, a barcode must be flexible in both data capacity and network connectivity. Consider, for example, a ringtone downloading service advertised in a magazine. If an index 2D barcode were used, the only option for a user would be to connect his camera phone to the designated Web page and download ringtones via the Internet. Continuous network connectivity has to be guaranteed for this type of service always to be available. Conversely, no network connectivity leads to no service.

With a database 2D barcode with high data capacity, however, it is possible to provide other options. For example, a user can save the uniform resource locators (URLs) of Web sites directly encoded in the barcode in his or her mobile camera phone for later use. If a sample ringtone or the actual ringtone itself is encoded in the 2D barcode, he or she can play it on the camera phone and download it directly from the 2D barcode symbol with no network connection.

Database 2D barcodes allow a much greater amount of data within a smaller space, as compared with their 1D counterparts. However, most commonly used database 2D barcodes for mobile applications (e.g. QR Code and Data Matrix) were initially developed for industrial use where dedicated scanners are used to read the barcode symbols. Such dedicated scanners are capable of producing high-fidelity images and consequently can decode small barcode symbols made up of very small cells. In contrast, the data capacity for mobile applications of these 2D barcodes is rather limited owing to the limitations of mobile phone cameras. This allows them to encode only a small amount of data such as URLs and personal details.

Thus, in order to achieve our goal in this prototype barcode development, our target was to develop a database 2D barcode with the following strengths:

(i) improved data capacity within a given fixed space;
(ii) tolerance of physical damage; and
(iii) robust and reliable operation for resource-limited mobile devices.

Considering the limitations of camera phone capability, the only way to achieve the first aim is to increase the data density without reducing the cell size. Increasing the number of cells is not an option. With a black and white encoding scheme, each colour represents only a single bit, either 1 or 0. However, it is possible to encode more than one bit within a single cell by using a colour encoding scheme, enabling each colour symbol to represent more than a single bit.

The second and third aims can be accomplished:

(i) by developing effective and reliable encoding and decoding algorithms; and
(ii) by adopting technologies and techniques, such as the error detecting and correcting codes and the other useful schemes discussed in Chapter 5, that can enhance the robustness of 2D barcodes.

Hence, our prototype 2D barcode will be 'a novel colour-database 2D barcode optimised for mobile applications', which, in consequence, can be used as a tool to realise ubiquitous computing.

There are colour 2D barcodes currently in use, such as ColorCode and High Capacity Color Barcode (HCCB). However, these barcodes only use a limited number of colours, say four, to generate a symbol for mobile application. Furthermore, some of these colour barcodes only encode a limited capacity of data, making them index 2D barcodes. Such 2D barcodes often use colour elements to create an eye-catching symbol rather than to increase data capacity.

In order to improve siginificantly the data capacity, or, more precisely, the data density of a barcode within a given fixed space, our challenge was to develop a colour 2D barcode that uses more colours than was previously possible for camera phones. This feature is the highlight and the novelty of our colour 2D barcode, formally called the Mobile Multi-Colour Composite (MMCC™) two-dimensional barcode.

6.2   Designing a 2D symbology

6.2.1   Symbol design

Developing a barcode symbology and its system involve two major tasks: symbol design and code design. Although these tasks are closely related, the former mainly concerns the physical design of the symbology (e.g. the symbol structure, finder pattern, symbol shape and cell shape) whereas the main focus of the latter is to develop the encoding algorithm for a barcode. Through the code design, the data capacity of a symbology, the error detecting and correcting code used, the barcode format and so forth are determined.

Symbol shape

Both the symbol and cell shapes of the MMCC™ code were designed with the aim of improving the data density within a given space. Since the display of a camera phone is usually square or rectangular, the shape of MMCC™ code is based on a square pattern so as to use the available space maximally. A square symbol shape can be more easily and space-effectively accommodated within the phone screen than for example a circular shape or a rectangular shape like that used in 1D barcodes.

The cell shapes of 2D barcodes vary. For example, hexagonal cells are used for encoding MaxiCode and BeeTaggs. High Capacity Color Barcode (HCCB) encodes triangular cells, claiming that they occupy less physical space than square cells. Indeed, we can double the data density by arranging an array of triangular cells in such a way that each square cell is divided equally by a diagonal separator (see Figure 6.1(a)). However, the base of each triangular cell of HCCB is approximately 1.5 times as long as its height (Figure 6.1(b)), taking more space for each cell than the arrangement illustrated in Figure 6.1(a). Furthermore, there are two extra half-sized triangular cells, one at the left end and the other at the right end in each row of an HCCB symbol, resulting in a waste of the available space within a symbol.

In order to read the colour values of each cell accurately and retrieve the correct data, it is essential to sample the centre pixel of each cell or a pixel that is as close to the centre as possible. The reason is that the centroid of each cell is furthermost from the neighbouring cells and consequently detection of it will be least affected by the colours of neighbouring cells. Thus, although arranging the triangular cells in the way presented on the left-hand side of Figure 6.1 can increase the data density, it makes it difficult for decoding devices to pinpoint the centroid of each cell.

Conforming to our aim of maximising data capacity, the cell shape of MMCC™ code was chosen to be square, which makes the most of the available space within a symbol.

(a) 2D barcode with triangular cells   (b) Triangular cells used for HCCB

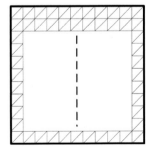
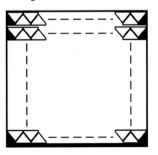

Fig. 6.1   Examples of 2D barcodes with triangular cells.

Square cells can be arranged in such a way that there is no waste space between cells or within a symbol. It also gives more redundancy to each cell, which may enhance accurate symbol reading. This, in turn, improves the reading robustness.

Designing the finder pattern

One of the most important components of the 2D barcode symbol is its finder pattern. Most finder patterns are used not only for detecting and locating barcode symbols but also for acquiring geometric information such as the size, shape, orientation and any distortion of the target symbol. The entire computation is performed using the geometric properties of the finder pattern. Furthermore, finder patterns often serve as a tool to correct the orientation and distortion once all the required information has been collected.

As discussed in Chapter 5, the size of the finder pattern (e.g. its height and width) decreases as the size of the cells is reduced in an effort to increase the data capacity of a 2D barcode within a given space. To ensure robust detection of the finder pattern, it needs to be designed in such a way that it is large enough to be accurately located in its smallest version and is capable of providing accurate geometric information so that the reading software can measure the target symbol precisely.

We examined different types of finder pattern currently in use in terms of usability, robustness and operation time. These included the discrete finder patterns found in Visual Code and QR Code, finder patterns with a combination of solid borders and broken borders like those adopted by Data Matrix and Trillcode, the continuous finder pattern or bounding border found in ColorCode and HCCB and the central circular finder pattern used in ShotCode and MaxiCode.

We also developed a prototype finder pattern and conducted experiments to investigate the capability, advantages and disadvantages of a particular discrete finder pattern [62]. The structure of the finder pattern is presented in Figure 6.2. It is made up of two major components: a *guide area* and a *reference area*. The guide area consists of two guide blocks and an L-shaped guide bar, whereas the reference area is composed of two chequered borders into which the colour reference cells are integrated.

The experimental results indicated that the finder pattern and the recognition algorithm developed to detect the finder pattern can operate robustly when the cell size of an interim colour 2D barcode and the two guide blocks were comparatively large. However, when locating sampled symbols having a greater data capacity (i.e. symbols whose cell size is small), the performance was not very impressive.

The novel feature of the prototype finder pattern was its integration with additional information such as colour reference cells. The experiment results demonstrated the feasibility of the idea. The finder pattern is the first object to be located by the reading software; thus, its successful location ensures the sampling of the colour reference cell values at their correct locations.

The three key findings of the experiment were as follows.

(i) It is very important to develop an effective thresholding or binarising method in order to locate a finder pattern robustly.

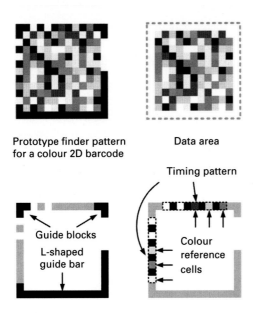

Prototype finder pattern   Data area
for a colour 2D barcode

Anatomy of finder pattern

Fig. 6.2   Structure of a prototype finder pattern. For the colour version, please see Plate 7.

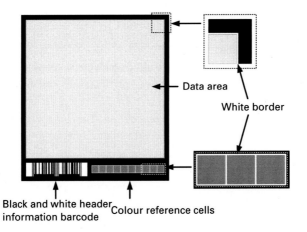

Fig. 6.3   Symbol structure of MMCC™.

(ii) Symbol distortion is one of the major causes for failures in measuring the geometric properties of a finder pattern.

(iii) A continuous, rather than a discrete, finder pattern having a reasonably large line-width (see Figure 6.3) may be a better choice for database 2D barcodes [62].

Database 2D barcode symbols are scalable in size. Depending on the data capacity and the space available for printing, the size of the components of a finder pattern may become very small. Poor thresholding often changes their sizes, more than likely from larger to smaller, preventing the reading software from locating them. Symbol distortion also changes the properties of a finder pattern such as its shape, size, orientation and the angles between components, which may result in either a false negative or a false positive. In the experiments, the key for successful symbol detection was correct location of the two guide blocks, especially when the data capacity was increased; the L-shape guide bar, whose selection criterion was based on its area, was always robustly detected. Hence, to minimise the chance of losing any component of the finder pattern completely, including all the components in one continuous frame may be a good idea. This may allow the reading software to work properly even when the finder pattern is partially damaged.

In view of the outcomes of our experiments, a solid border was adopted for our prototype 2D barcode since it fulfils the requirements for the MMCC™ barcode to maximise the data capacity of the available space while ensuring robust reading of the symbols. The novelty of our new finder pattern is that it has additional information pertaining to the barcode, such as the barcode version and error correction level as well as a colour reference area, within the finder pattern. The format information is first encoded into either a black and white 1D barcode or a black and white 2D barcode, which is then embedded into the finder pattern, along with the reference cells for colour calibration.

The format information and colour reference cells are usually included within the data area of a symbol. However, this not only reduces the space for encoding data but also complicates the structure of a 2D barcode, which may invite unnecessary errors. However, MMCC™ barcode strengthens its robustness of operation by acquiring the header information from the black and white barcode embedded in the finder pattern since the recognition of black and white can be less troublesome and more accurate than the recognition of colours.

The header information can be encoded into either a 1D barcode or a 2D barcode format, depending on the amount of data required. It can include not only information such as the barcode version and error correction level, as mentioned earlier, but also security settings for confidential data and parameters for special functions. This allows MMCC™ barcode to be a self-contained secure container and also to provide flexible operation of the MMCC™ barcode system, according to the user requirements. Figure 6.3 illustrates the symbol structure of MMCC™ barcode.

Figure 6.3 is an example of an MMCC™ barcode symbol where a 1D barcode is embedded to provide the header information. The area for the colour reference cells is adjacent to the 1D barcode. Each colour reference cell is surrounded by a four-pixel-wide white border to minimise the effect of neighbouring colours and of the black border of the finder pattern. The same applies to the colour cells of the data area. A pixel-wide white border separates the data area of the MMCC™ barcode from the black finder pattern.

6.2.2 Colour selection for encoding data

The data capacity of a colour 2D symbology is subject to how many colours can be robustly retrieved after the symbol has been through the various colour channels (usually called *colour spaces* or *colour models*[1]) and compression and decompression of the symbol's image. For example, if four colours are available for encoding data, each cell can have one colour from the four, each colour representing two-bit data, 0 0, 0 1, 1 0 or 1 1. Similarly, using eight colours three-bit data can be represented by each cell, which results in an increase in data density for a given space without a change in the number of cells.

To ensure the robustness of barcode reading, colours that are furthest apart in a particular colour space are usually selected for encoding data into colour cells. However, most barcode systems involve more than one colour space in their operation process, which necessitates a colour conversion between the different colour spaces. For example, a barcode symbol generated using one colour space by hardware such as computers or mobile devices is usually printed with a printer that uses another colour space. This means that two colour spaces have already been used to generate a barcode symbol printed on materials such as paper. The number of colour spaces involved in a barcode system might be reduced if the system only involves barcode symbols displayed on the monitors of hardware. However, such systems considerably limit the usability of the barcode. Hence, it is important to select colours that can preserve their distances from one another after undergoing conversion between the different colour spaces. This urged us to address the effect of such colour conversions on colour values.

In addition to the conversions between colour spaces, there are other factors that affect the colour values of barcode symbols during image capturing and processing. Lighting is the primary factor we need to consider. A *gamma correction* or *white balance* might be automatically performed by the built-in camera of a mobile phone to minimise the lighting effect and produce better images. What is actually performed by these cameras greatly depends on the CCD or CMOS sensor used [63]. Whether or not it is favourable, there is certainly an influence on the values of the resulting colours.

The file formats employed for storing captured images also have a significant impact on the fidelity of the reconstructed images, because of the compression and decompression algorithm involved in the formatting process. The deterioration in image fidelity is remarkable, especially when a lossy image compression algorithm such as that of the Joint Photographic Experts Group (JPEG) is used.

Before entering into the colour selection process, these factors should be addressed in detail to ensure an optimal selection of colours. Hence, the fundamentals of colour and colour spaces, the impact of lighting on the colour values and the effect of file formatting on the fidelity of a reconstructed image need to be considered in advance.

Fundamentals of colour

Basically, the colour perceived by humans in an object is determined by the nature of the light reflected from the object. That is, colour vision is enabled by the presence of three elements: a light source, an object and a human eye as a detector. Without light, colours do not exist. Hence, the characterisation of light is central to the science of colour [64].

---

[1] Strictly, in colour science, the term *colour space* is restricted to quantitative systems that can be mathematically transformed. Any other colour systems should be referred to as *colour models*.

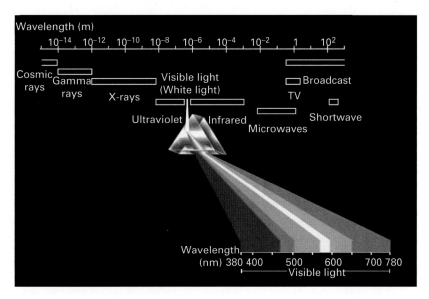

Fig. 6.4   Colour spectrum (adapted from [65]). For a colour version, please see Plate 8.

When a beam of sunlight passes through a glass prism, the emerging beam of visible light consists of a continuous spectrum of colours that includes six broad regions: violet, blue, green, yellow, orange and red.[2] Differences in wavelength are perceived by the human eye as different colours, and result in a colour spectrum ranging from violet at the shortest visible wavelength to red at the longest. Thus visible light, which is part of the electromagnetic spectrum, is composed of a relatively narrow band of frequencies, ranging in wavelength from about 380 to 780 nanometres (nm). Electromagnetic radiation at wavelengths shorter than the violet end of visible light is *ultraviolet* and at wavelengths longer than the red end is *infrared*. Both ultraviolet and infrared are invisible. The combination of all the wavelengths of visible light appears white. Figure 6.4 shows the colour spectrum produced by dispersing white light by a glass prism.

The human eye perceives the colour of images using cone-shaped light receptors lining the surface of the *retina*. There are between six and seven million such cones in each eye. As a result of detailed experiments, it has been proved that the cones are sensitive to three distinct but broadly overlapping regions of the light spectrum, corresponding to red (R), green (G) and blue (B) [64]. Approximately 65% of all cones are sensitive to red light and 33% are sensitive to green light. The cones sensitive to blue light account for only about 2% of all, although their sensitivity is the highest.

Conceptually, all the colours perceived by human eyes can be represented by combinations of these three colours. Hence, R, G and B are called the *primary colours (of light)*. In 1931, for the purpose of standardisation, Commission Internationale de l'Eclairage (the International Commission on Illumination or CIE) defined the primary colours to

---

[2] Traditionally, there are seven colours, 'indigo' being included between violet and blue.

have the specific wavelengths 435.8 nm (blue), 546.1 nm (green) and 700 nm (red).[3] In practice, however, no single colour value can represent the colours red, green and blue. Furthermore, and more importantly, it has been found that some colours cannot be produced by a combination of these primary colours [64], which urged the CIE to develop one of the first mathematically defined colour spaces, the *CIE 1931 XYZ colour space* (also known as the *CIE 1931 colour space*), in which all visible colours can be represented.

Colour space
RGB colour space

The RGB colour space, which stems from the three primary colours of light, is an additive colour space where a wide range of colours can be produced by adding various combinations of the primary colours. Mixing the same amount of the primary colours in the correct intensities produces white light.

As presented in Figure 6.5 (see also Figure 4.17(a)), the RGB colour space can be represented as a three-dimensional cube where red, green and blue are located at the corners of each axis. Cyan, magenta and yellow are at the other three corners, between green and blue, red and blue and red and green, respectively. Black is at the origin and white is at the diagonally opposite corner of the cube. A greyscale, shades of grey produced by equal values of red, green and blue, forms the diagonal line joining the black and white points. In a 24-bit colour graphic system, where each channel has eight bits per colour, the value of each primary colour is assumed to be in the range [0, 255]

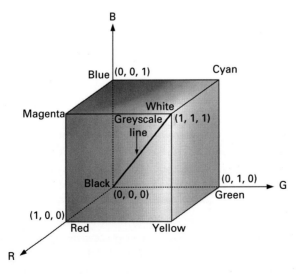

Fig. 6.5   RGB colour cube. For a colour version, please see Plate 9.

---

[3] The CIE standard values were set before detailed studies on the specific wavelength values of the primary colours and are slightly different from the experimental data that became available in 1965.

(see Section 4.3.2). For the colour cube, the colour values of red, green and blue are assumed to have been normalised to be in the range between [0, 1].

The RGB colour space does not define the exact values of red, green and blue. As a result, sharing a common colour space does not necessarily guarantee that the identical colour can be reproduced by defining a colour with an identical combination of red, green and blue colour values. The colours actually produced vary noticeably depending on the devices employed [66]; hence, the RGB colour space is known as a device-dependent colour space.

CMY or CMYK colour space

By adding the primary colours of light, the secondary colours of light, namely, cyan, magenta and yellow, can be produced (see Figure 6.5).

Whereas monitors emit colours as RGB light, inked paper absorbs or reflects specific wavelengths of light. For example, when the surface of paper coated with cyan pigment is illuminated with white light, only green and blue light are reflected from the surface; red light has been subtracted from the reflected white light. The resulting colour is cyan. The primary colours of pigments or colourants, consequently, are defined as those that subtract or absorb a primary colour of light and reflect or transmit the other two [64]. Hence, a colour space based on the primary colours of pigments, cyan, magenta and yellow, is called a subtractive colour model or space. Figure 6.6 presents for comparison the RGB additive colour space and the CMY subtractive colour space. Cyan, magenta and yellow, which are the secondary colours of light, are the primary colours of pigments. The colours of pigments serve as filters, subtracting varying degrees of red, green and blue from white light so as to produce a selective *gamut* (range) of spectral colours.

In an additive colour space such as RGB, white is produced by adding equal amounts of the primary colours of light, whereas the absence of light results in black. In a subtractive colour space such as CMY, the resulting colour of a proper combination of the three

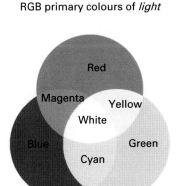

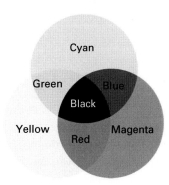

Additive colour mixture of RGB primary colours of *light*

Subtractive colour mixture of CMY primary colours of *pigment*

Fig. 6.6   RGB additive colour space and CMY subtractive colour space. For a colour version, please see Plate 10.

pigment primaries should be black. However, in practice, the result is brown or grey due to the turbidity of pigments, the transparency of inks or so forth. In addition to cyan, magneta and yellow, black ink is normally used to produce deeper black tones and also to reduce printing cost. The colour space obtained by adding black (K) is called CMYK.

On the one hand, most devices that deposit coloured pigments on paper (e.g. colour printers and copiers) require CMY or CMYK data input. On the other hand, RGB data are processed by devices that sense, represent and display images on electronic systems such as digital cameras, scanners, televisions and the displays of computers or camera phones. Consequently, when CMY(K) data are not initially available, an RGB to CMY conversion must be performed. The conversion is usually performed internally, according to the following vector equation:

$$\begin{bmatrix} C \\ M \\ Y \end{bmatrix} = \begin{bmatrix} 1 \\ 1 \\ 1 \end{bmatrix} - \begin{bmatrix} R \\ G \\ B \end{bmatrix}, \tag{6.1}$$

where the assumption is that all colour values have been normalised to the range [0, 1] [64]. Equation 6.1 is based on the idea that a pigment primary subtracts or absorbs a primary colour of light and reflects or transmits the other two. For example, the CMY data of pure cyan can be acquired from the equation $C = 1 - R$.

When performing an RGB to CMY conversion, however, there are two major obstacles that make it difficult to accomplish the accurate conversion of colour values: the device dependence of both the RGB and the CMY(K) colour spaces and the difference in colour gamut between the two colour spaces. The device-dependent nature of RGB space, resulting from ambiguous definition of the standard red, green and blue values, causes the same RGB values to be reproduced as noticeably different colours on different devices [67]. To make matters worse, CMY(K) colour space is also device-dependent. It is obvious that accurate RGB to CMY conversion can hardly be performed if the specified colour values may vary even when sharing a common colour space.

The difference in colour gamut between the RGB and CMY colour spaces also affects considerably the accuracy of colour conversion between them. To visualise gamuts as areas in order to show the difference in gamuts, the CIE 1931 chromaticity diagram is often used since it may be easier to see the difference in a 2D space rather than in a 3D space.[4] As Figure 6.7 demonstrates, the colour gamut produced by RGB monitors is wider than the gamut achieved by colour printing devices that use CMY(K) colour data. It is obvious that the printing devices cannot reproduce some colours that are within the gamut of RGB monitors. In consequence, there is no simple or general conversion formula that can ensure accurate colour conversion between RGB and CMY(K) spaces.

As a solution in a professional environment, *colour matching* is performed for all the devices involved through a colour management system that aims to accomplish accurate colour conversions between devices, using the colour profile attached to each

---

[4] In effect, the brightness or value should also be added to represent the full gamut. Accordingly, the full gamut should be represented in 3D space.

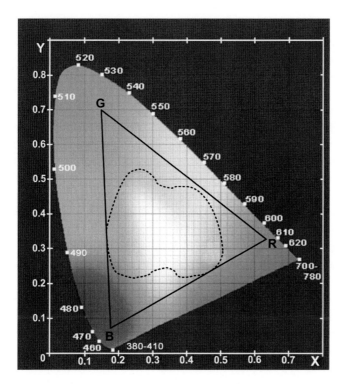

Fig. 6.7 Colour gamuts of colour monitors (solid black line) and colour printing devices (broken black line), for comparison. The mainly pale grey horseshoe corresponds to the visible range of the electromagnetic spectrum (adapted from [64]). For a colour version, please see Plate 11.

device. A colour profile known as the *ICC profile* describes the colour-space range produced by each device, following the format promulgated by the International Colour Consortium (ICC). Colour management may minimise the damage to colour accuracy introduced by colour conversion between different devices. Its effect can be significant, especially when the devices involved in the conversion have different gamuts. Nevertheless, colour management cannot avoid such damage completely; the accuracy in colour value is bound to be reduced after RGB to CMY(K) colour conversion.

YCbCr colour space

The YCbCr colour space family is used in video systems; Y is the *luminance* (i.e. brightness) component and Cb and Cr are the blue-difference and red-difference *chrominance* (i.e. colour information) components [67].

The YCbCr colour space was developed as part of ITU-R BT.601[5] during the development of a worldwide digital-component video standard.[6] All luminance–chrominance

---

[5] A standard published by the International Telecommunications Union – Radiocommunications sector for encoding interlaced analogue video signals in digital form.

[6] When this model is used for analogue component video, YCbCr is often called YPbPr, although the term YCbCr is often used for both systems.

formats used in the various television (TV) and video standards, such as YIQ for the National Television Systems Committee (NTSC) format and YUV for the phase alternation line (PAL) format, use colour difference signals, by which RGB colour images can be encoded for broadcasting or recording and later decoded into RGB again to be displayed. These intermediate formats were needed for compatibility with the pre-existing black and white TV formats.

The use of RGB space simplifies the design of computer graphics systems and so this space has been widely used. However, the red, green and blue components are highly *correlated* in the RGB colour space, which makes it difficult to implement some applications where only a particular colour component is of interest. In YCbCr colour space, however, the luminance and chrominance components can be separated, allowing each component to be manipulated separately. This feature was required for compatibility with the pre-existing black and white TV formats. It is also useful for many image processing techniques, such as histogram equalisation, that only work on the intensity component of an image [67]. Furthermore, colour difference signals such as YCbCr require a lower data bandwidth than full RGB signals.

The human eye is less sensitive to fine colour details than to fine brightness details. This allows the removal of a certain amount of colour information without its being noticed, which makes it possible to achieve a high image data compression ratio. There are several YCbCr sampling or *subsampling* (also known as *downsampling*) formats such as 4 : 4 : 4, 4 : 2 : 2, 4 : 1 : 1 or 4 : 2 : 0 ratios. Subsampling results in lossy compression, since data that are not sampled are lost permanently. Current high-efficiency digital-colour-image data compression schemes such as JPEG and MPEG (i.e. Moving Picture Experts Group) store the RGB colours internally in YCbCr format or in a digital luminance–chrominance format.

According to the ITU-R BT.601 standard, the conversion between RGB and YCbCr can be computed as follows:

$$\begin{bmatrix} Y \\ Cb \\ Cr \end{bmatrix} = \begin{bmatrix} 0.299 & 0.587 & 0.114 \\ -0.169 & -0.331 & 0.5 \\ 0.5 & -0.418 & -0.0813 \end{bmatrix} \begin{bmatrix} R \\ G \\ B \end{bmatrix} + \begin{bmatrix} 0 \\ 128 \\ 128 \end{bmatrix},$$

$$\begin{bmatrix} R \\ G \\ B \end{bmatrix} = \begin{bmatrix} 1 & 0 & 1.402 \\ 1 & -0.344 & -0.714 \\ 1 & 1.772 & 0 \end{bmatrix} \begin{bmatrix} Y \\ Cb - 128 \\ Cr - 128 \end{bmatrix}.$$

The standard also defines Y to have a nominal eight-bit range 16–235 and Cb and Cr to have the range 16–240 (see Table 6.1), which correspond to the RGB range 0–255. However, owing to noise, occasionally Y and CbCr go outside their respective colour ranges, 16–235 and 16–240.

The above equations define YCbCr colour space in such a way that the rotated RGB colour cube fits within a larger YCbCr colour cube. Figure 6.8 illustrates the relation between the RGB and YCbCr colour spaces. From this figure, it is obvious that there are YCbCr colour values outside the valid RGB values. This causes difficulty in determining

Table 6.1. Range of the components of RGB and YCbCr

Colour space	RGB			YCbCr		
Component	R	G	B	Y	Cb	Cr
Min	0	0	0	16	16	16
Max	255	255	255	235	240	240

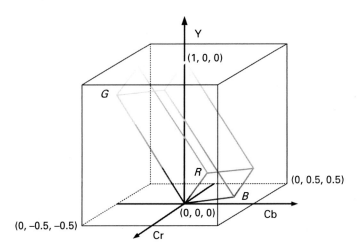

Fig. 6.8   Relation between RGB and YCbCr colour space (adapted from [67]). For a colour version, please see Plate 12.

how to interpret and display YCbCr colour values outside the valid RGB range correctly, resulting in inaccurate colour conversion.

The YCbCr system is adopted in video codec and transmission, whereas RGB is adopted for display; thus, conversion between them is unavoidable [68]. Hence, it is important to ensure accuracy in the data throughout the colour conversion. However, similarly to the colour conversion between RGB and CMY(K), the conversion between RGB colour space and YCbCr colour space has a negative effect, to a certain degree, on the accuracy and consistency of the colour data even when the colour management between them is performed accurately. That is, YCbCr to RGB conversion with precision reduction is a prerequisite in many applications which involve it [68].

HSI, HLS, HSV and HSB colour spaces

The HSI (hue, saturation, intensity), HLS (hue, lightness or luminance, saturation), HSV (hue, saturation, value) and HSB (hue, saturation, brightness) colour spaces were developed to be more 'intuitive' in manipulating colour and were designed to approximate the way in which humans perceive and interpret colour. A brief description of the HSB and HSV colour spaces was provided in Chapter 4. However, these colour spaces are not involved in our barcode operation process and therefore their detailed description is beyond the scope of this book.

JPEG file formats

The fidelity of the image reconstructed by a camera phone varies according to the file format used to store the image, since different file formats use different compression and decompression schemes. It is possible to develop a barcode decoding algorithm targeting an uncompressed two-dimensional array. In fact, this is computationally more efficient, saving memory and CPU time for both compression and decompression. However, it may limit the possible applications since it eliminates the option of storing the captured images. Furthermore, a decoding algorithm developed based on the best possible image may not work for images with less quality. In fact it is more likely that a decoding algorithm can cover the whole range of images if it is based on the images with the least fidelity.

There are various file formats to store the images captured by a camera phone: JPEG, tagged image format file (TIFF), bitmap (BMP), graphic image format (GIF), portable network graphics (PNG) and so forth. Among them, JPEG is one of the most popular still-frame compression standards, owing to its high level of compression capability with little perceptible loss in image quality. Most photographic image capturing devices such as digital cameras use the JPEG file format [63].

The compression techniques can be classified into two broad categories: *lossless compression* (also known as *information preserving compression*) and *lossy compression* [64]. The former allows an image to be compressed and decompressed without any loss of information. However, the original image cannot be retrieved with the latter compression technique. The advantage of the lossy compression technique is that it can achieve high levels of data reduction while preserving a satisfactory image quality, which often deceives humans into believing that the image is the original [63].

The JPEG format defines three different coding systems: a lossy *baseline coding system*, an *extended coding system* for greater compression or higher precision and a lossless *independent coding system* for reversible compression [64]. Among them, the baseline coding system is the best-known implementation and is based on a discrete cosine transform (DCT) compression of 8×8 blocks of pixels [69].

Although lossless compression with it is possible, JPEG compression is generally known as a lossy compression technique since lossless JPEG compression has been used only rarely. The widespread use of lossy JPEG compression demonstrates that the perceivable loss in image quality is little. However, this does not change the fact that part of the data have been lost permanently. Hence developing a barcode decoding algorithm based on the JPEG image format is rather challenging. It seemed nevertheless reasonable to select the JPEG format as our target image compression format, considering that most camera phones use this format. Furthermore, a decoding algorithm that works well for images with low fidelity should be able to handle images with better quality.

JPEG compression algorithm

To cope well with JPEG images, a clear understanding of its format is essential. However, the proposed barcode system does not involve the JPEG encoding and decoding processes but rather the resulting JPEG images. Hence, the JPEG compression algorithm will be explained with the focus more on the operations applied to an input image and why and how they affect the resulting image (i.e. what is preserved and what is lost), rather than a detailed step-by-step description.

The Joint Photographic Experts Group is, in fact, the name of the committee that created the standard, and no particular file format, spatial resolution or colour space model is specified for the JPEG format [64]. However, *JPEG File Interchange Format* (JFIF) encoding is considered to be a de facto standard format. The JPEG baseline coding system with JFIF compression process involves the following steps.

(i) The colour space is converted from RGB to YCbCr, wherein the Cb and Cr components are subsampled (see Figure 6.9).
(ii) Each component of the image is divided into blocks of 8×8 pixels.

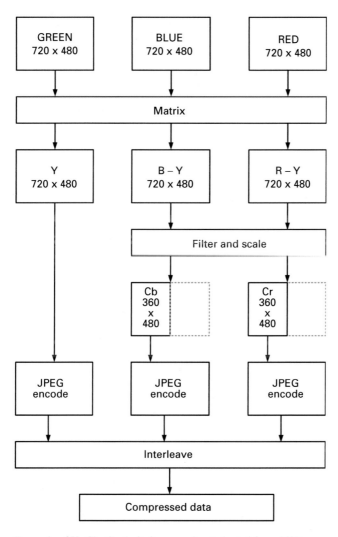

Fig. 6.9 Example of $Y : Cb : Cr$, $4 : 2 : 2$ processing (adapted from [69]).

(iii) A DCT is performed to convert each 8×8 block to a frequency-domain representation.

(iv) *Quantisation* is performed.

(v) The coefficients are scanned in a diagonal zigzag order.

(vi) *Entropy coding* is performed.

The JPEG format achieves large compression ratios by two colour-information-removal steps, subsampling and quantisation (which rounds off the DCT coefficients).

These operations reflect the *human visual system* (HVS). The subsampling operation builds on the fact that the HVS is less sensitive to colour details (i.e. chrominance) than to brightness details (i.e. luminance). The quantisation operation builds on another feature of the HVS, that the human eye is very sensitive to low-frequency coefficients but much less sensitive to amplitude errors in the higher-frequency coefficients [69]. This enables coarser quantisation for the higher-frequency coefficients. Thus a reduction in the chrominance data and high-frequency coefficients can be made without affecting the human eye's perception.

This explains why a YCbCr colour space is used for JPEG compression. The values of each R, G and B component are of equal importance. Furthermore, in RGB colour space the luminance and chrominance components are not separated, which makes it difficult to subsample only the chrominance components. However, these components are separated in YCbCr colour space, in which Y represents luminance and both Cb and Cr represent chrominance. This allows separate operations for each component. Furthermore, only the luminance operation is required for black and white images.

As explained previously, there are several YCbCr subsampling formats such as $4:2:2$ (see Figure 6.9) and $4:2:0$. This subsampling is a factor leading to lossiness in JPEG compression. After this step, each component of YCbCr colour space, which is processed separately in the subsequent process, is divided into 8×8 blocks for better performance. It is then converted to a frequency-domain representation by DCT, yielding 64 coefficients. The DCT coefficient at a location (u, v) can be computed from each of the 64 pixels at locations (x, y) as

$$F(u, v) = \frac{1}{4} C_u C_v \sum_{x=0}^{7} \sum_{y=0}^{7} f(x, y) \cos\left(\frac{(2x+1)u\pi}{16}\right) \cos\left(\frac{(2y+1)v\pi}{16}\right), \quad (6.2)$$

$$C_u, C_v = \begin{cases} \frac{1}{\sqrt{2}} & \text{for } u, v = 0, \\ 1 & \text{otherwise.} \end{cases}$$

Figure 6.10 shows a sample image (top left), its 8×8 pixel block (bottom left), the colour values of the pixels (top right) and the resultant values of 128 subtraction (bottom right). Figure 6.11 shows the resulting coefficients of DCT operation and the quantisation matrix for the next step.

The top left entry in Figure 6.11(a) is called the *DC coefficient*. The remaining 63 coefficients are called the *AC coefficients*. For an eight-bit image each pixel has 256 possible values, in the range [0, 255]. From Equation 6.2, consequently, $f(x, y)$ in this

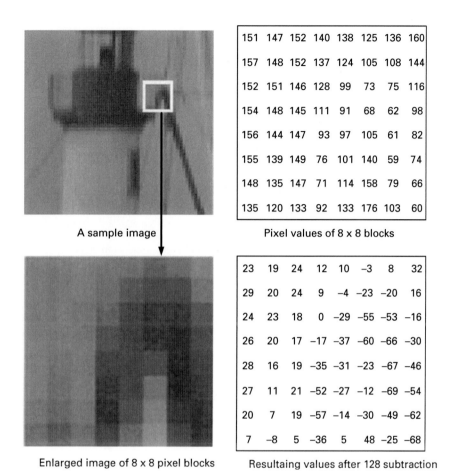

151	147	152	140	138	125	136	160
157	148	152	137	124	105	108	144
152	151	146	128	99	73	75	116
154	148	145	111	91	68	62	98
156	144	147	93	97	105	61	82
155	139	149	76	101	140	59	74
148	135	147	71	114	158	79	66
135	120	133	92	133	176	103	60

A sample image   Pixel values of 8 x 8 blocks

23	19	24	12	10	−3	8	32
29	20	24	9	−4	−23	−20	16
24	23	18	0	−29	−55	−53	−16
26	20	17	−17	−37	−60	−66	−30
28	16	19	−35	−31	−23	−67	−46
27	11	21	−52	−27	−12	−69	−54
20	7	19	−57	−14	−30	−49	−62
7	−8	5	−36	5	48	−25	−68

Enlarged image of 8 x 8 pixel blocks   Resultaing values after 128 subtraction

Fig. 6.10   A sample image, its enlarged 8×8 pixel block and pre-operations to performing DCT (adapted from [69]).

range yields AC coefficients in the range -1023 to $+1023$, which can be represented by an 11-bit signed integer. However, the DC coefficient would be in the range [0, 2040] and would be represented by an unsigned 11-bit integer, which would require different hardware and software implementations. To avoid this, the value 128 (half the pixel range of an eight-bit image) is subtracted from every pixel value prior to the DCT operation [69].

These coefficients are the subject for the quantisation operation. A great deal of information in the high-frequency domain is reduced in this step, making the most of the fact that the human eye is rather insensitive to brightness variations of high frequency as compared with small differences in brightness over a relatively large area (i.e. low-frequency brightness variations). This is another factor in the lossiness of JPEG in the whole compression operation. Quantisation is performed by simply dividing each DCT coefficient by the corresponding value from the quantisation matrix (see Figure 6.11).

(a) The 64 coefficients (1 DC and 63 AC) yielded by the DCT operation

-63	157	22	11	-23	-37	68	11
62	-17	37	-90	64	15	-46	-2
66	-71	-25	37	-17	8	-2	4
16	-7	8	14	-14	-11	20	3
5	-12	-7	2	3	8	-12	2
-7	6	7	-7	-2	-3	-2	3
-3	-1	-4	0	1	1	0	-5
0	1	-1	0	-3	0	-2	0

(b) The quantisation matrix

16	11	10	16	24	40	51	61
12	12	14	19	26	58	60	55
14	13	16	24	40	57	69	56
14	17	22	29	51	87	80	62
18	22	37	56	68	109	103	77
24	35	55	64	81	104	113	92
49	64	78	87	103	121	120	101
72	92	95	98	112	100	103	99

Note: Values more than 95 are shaded.

Fig. 6.11   (a) The 64-coefficient yield of the DCT operation and (b) the quantisation matrix for the luminance; values more than 95 are shaded in grey (adapted from [69]).

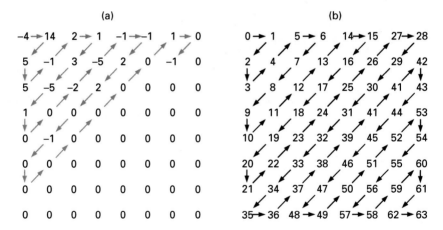

Fig. 6.12   (a) Quantisation result and (b) diagonal zigzag scanning for the entropy coding (adapted from [69]).

Figure 6.11(b) shows minima, at (0, 2) and (1, 0), which correspond to the maximum of the luminance-contrast sensitivity curve for the HVS [69], whereas the area with the highest values, shaded in grey, corresponds to the DCT coefficients in the high-frequency domain. Division by these large values causes the high-frequency coefficients to be rounded to zero. The advantage of the DCT is its tendency to concentrate energy into the top left coefficients, which is accentuated in the quantisation step that follows [69]. The quantisation result indicates that the probability of a coefficient's being non-zero is much higher in the top left than at the bottom right. This probability is likely to follow a diagonal zigzag scanning pattern (see Figure 6.12(b)), which is why this scanning scheme is adopted: it not only reduces the sizes of the DCT coefficients as a whole but also increases the efficiency of the subsequent entropy coding.

The scanning output is further compressed by entropy coding based on Huffman coding. Since the latter is a lossless algorithm, there is no further negative effect on the quality of the target image. The decoding process is simply the reverse of the encoding steps.

Rather than explaining the entropy coding and JPEG decompression in detail, the key features of the lossy JPEG compression and its effect on the robustness of colour 2D barcode reading have been provided here.

Key features of JPEG compression

As mentioned above, the key features of JPEG that affect colour 2D barcode reading are:

(i) the colour information is reduced owing to the subsampling of the chrominance component;

(ii) a great deal of information in the high-frequency domain, where the pixel values change rapidly, is removed, which often introduces a blurring effect.

It is clear that a JPEG compression is suited to continuous-tone images (e.g. photographic and real-world images) but not to discontinuous images (e.g. colour 2D barcode). Hence, developing a colour 2D barcode decoding algorithm targeting JPEG images is another factor that makes colour 2D barcode reading a rather challenging task.

Lighting effect on colour values

The effect of the lighting arrangement on colour values is tremendous. To build up a better picture of how such factors affect the colour values, consider the image presented in Figure 6.13. This image consists of four areas, A, B, C and D. The question arises, 'Is B the same colour as A?' The following may be a typical human interpretation.

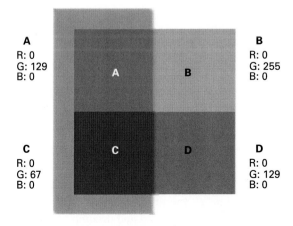

A
R: 0
G: 129
B: 0

B
R: 0
G: 255
B: 0

C
R: 0
G: 67
B: 0

D
R: 0
G: 129
B: 0

Fig. 6.13   Effect of lighting on colour values. For a colour version, please see Plate 13.

There are two strips AB and CD, each in uniform lightness and chrominance. Owing to a shadow cast on A and C, these colours are darkened. Nevertheless, A and B were originally the same colour, thus, we say that B is really the same colour as A.

Unlike subjective human vision, which compensates for varying brightness, image processing devices such as a camera phone capture and interpret an image as it is. When seen by these devices, A and B in Figure 6.13 are surfaces of different brightness and saturation [70], and so are C and D. This indicates that A and B are not identical colours, which is supported by the RGB colour values of each area presented on the left. In fact, the green value of A (i.e. 129) is only around a half of that of B (i.e. 255). Similarly, the green value of C (i.e. 67) is around a half of that of D. Remarkably, the areas A and D are recorded as the identical colours (i.e. $R = 0$, $G = 129$ and $B = 0$), having the same hue, saturation and brightness, although they might be perceived by the human eye as quite different colours. Figure 6.13 demonstrates the susceptibility of colours to the lighting effect. Identical colours of objects may become different colours, and vice versa.

When we talk about the lighting effect, we need to be aware that there are different types of lights, each having a different lighting effect. Sunlight is no exception. This means that there is always a certain amount of lighting effect even when no additional light is used. These light sources are not necessarily white; they do not necessarily have the full gamut of colours. That is, often a light source itself has a colour. Coloured lighting tends to neutralise and darken complementary-coloured surfaces, thus tending to raise the relative lightness of all surfaces that reflect wavelengths present in that light [70].

In order to minimise such a detrimental lighting effect and, as a result, to enhance robustness in barcode reading, either global thresholding or adaptive thresholding is often applied to the target image; this is discussed below.

Possible solutions to preserve accuracy in colour values

We have discussed the three major factors that have detrimental effects on the accuracy and consistency in colour values during barcode encoding and decoding operations. These include:

(i) colour conversion between different colour spaces;
(ii) a file format with lossy compression; and
(iii) the lighting effect.

To overcome these problems, effective measures need to be applied. As was discussed in Chapter 5, the use of a colour reference palette is a practical solution.

Use of reference colours

Colour reference cells provide the standard colour for each corresponding colour symbol used to encode data. The value of each colour cell in the data area is determined relative to the value of the corresponding standard colour in the reference cell. Since the relative difference in colour value of a data cell from that of the standard colour is consistent, no matter which colour conversions, compression and decompression or lighting have been applied to a colour 2D barcode symbol, the data can be retrieved correctly.

The use of reference colours for colour calibration is essential to read colour barcodes successfully and should be adopted as a first step to improve their reading robustness. However, this alone cannot solve all known problems. Two remaining problems occur:

(i)   when a colour 2D barcode symbol that was unevenly lit needs to be decoded; and
(ii) when several reference cells have similar colour values and, as a result, can be confused with each other.

With regard to the first of these problems, lighting does not always affect the entire symbol in a systematic way. When a colour reference cell and its corresponding data cell are under different lighting effects, it is more likely that their colour values will become quite different (see Figure 6.13), making it difficult to use the reference colour values validly. As we have seen, it is possible that the values of colours that were originally different are now similar or even the same. The probability of this is low when the values of these colours are as far apart as possible (e.g. black and white) within a colour space such as the RGB space. However, the reverse may be true when the distance between the colours is small.

One possible solution is to remove the lighting effect as much as possible. Hence adaptive thresholding, a well-known technique to minimise lighting effects, is worthy of investigation.

Thresholding to remove lighting effect

The images reproduced by image processing devices such as a camera phone are either greyscaled or coloured. Thresholding is a technique used for binarising an image. Each grey pixel is examined; the lighter pixels are made white and the darker pixels are made black [71]. Although there are many ways to threshold an image, thresholding methods can be classified broadly as *global thresholding* and *adaptive thresholding*.

Global thresholding binarises an image as explained above, setting all pixels below a certain grey level to black and all others to white. Hence, it is important to select the threshold grey level appropriately. Not only the range of actual pixels but also their distribution is often considered for selection of the threshold. The advantage of global thresholding is its simple and fast operation. However, global thresholding often fails to produce good results when the range of brightness varies greatly across an image [71]. Since a single global threshold value is used for the entire image, it is possible that some parts of the image will be incorrectly binarised (i.e. there will be too much black and too little white or vice versa).

When the target image is unevenly lit, adaptive thresholding is more accurate. In this technique the threshold value across the image is varied according to the background illumination at each pixel; an ideal adaptive thresholding algorithm would produce the same result when applied to an unevenly lit image as a global thresholding algorithm would produce when applied to a perfectly evenly lit image [71]. The brightness of each pixel is normalised to compensate for its being more or less illuminated and then a decision is made as to whether the result should be black or white. To determine the background illumination at each point, a blank image is first examined and used as a reference and then the pixel value in this reference image is subtracted from the corresponding pixel [71]. The result is then thresholded.

In practice, the brightness of a camera image is not constant, so the purpose of using thresholding in such a case is to remove the uneven-light effect. Hence an adaptive thresholding method should be the right choice.

We adopted the modified adaptive thresholding method described by [17] and originally developed by [71], where the basic idea is to use a weighted moving average of the grey values while running through the image in a snakelike fashion (i.e. alternating left-to-right and right-to-left scanline traversal). The direction of the background lighting of the image affects the outcome of adaptive thresholding. By scanning the image in a snakelike fashion, however, the effect can be reduced [71]. Nevertheless this causes another problem, an 'every-other-line effect': in the grey regions of an unevenly lit image, scanning in one direction produces the opposite result from scanning in the other direction (see Figure G.1 in Appendix G).

In the algorithm developed by [17], for each examined scanline the previous scanline is taken into account to compensate artefacts resulting from the zigzag traversal of the scanlines. The average $g_s(n)$ is updated according to

$$g_s(n) = g_s(n-1)\left(1 - \frac{1}{s}\right) + p_n,$$

where p_n denotes the grey value of the current pixel and $s = 1/8$ is the moving average width of the image; g_s is initialised at $g_s(0) = \frac{1}{2}cs$, where c is the maximum possible grey value. The colour of the thresholded pixel $T(n)$ is computed as ($t = 15$)

$$T(n) = \begin{cases} 1 & \text{if } p_n < \dfrac{g_s(n)}{s}\dfrac{100 - t}{100}, \\ 0 & \text{otherwise.} \end{cases}$$

In an experiment conducted to address the effect of this adaptive thresholding, we used the colour data image integrated in the prototype finder pattern (see Figure 6.2) as a sample. Since our prototype 2D barcode encodes data into colour cells, the adaptive thresholding method was applied to the three colour channels R, G and B separately, resulting in three binarised images. They were then combined together. Figure 6.14 presents the original colour symbol (top left), its image reconstructed by a Nokia 6600 camera phone (centre left) and the thresholded image (bottom left). It also presents the three colour channels after adaptive thresholding was applied (right).

The image reconstructed by the Nokia 6600 camera phone, especially the outer bottom of its background, is slightly greenish due to the lighting effect. However, the main part of the background colour of the thresholded image (bottom left) appears to be pure white, with RGB value (1, 1, 1) (in an assumed RGB range [0, 1]) (the measurement was done using the MATLAB software). The values of most colour cells in the data area suggest that the original colours have been retrieved. This demonstrates the strong capability of the adaptive thresholding method in removing the lighting effect.

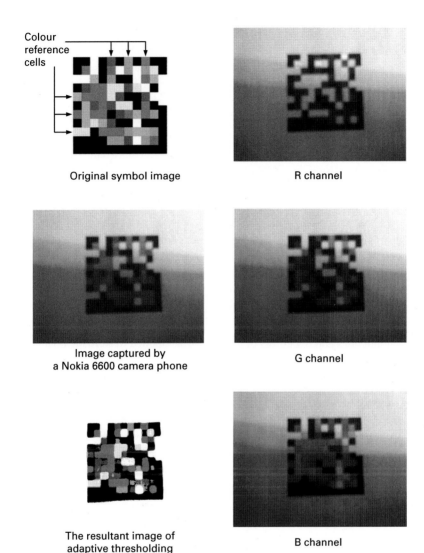

Original symbol image

R channel

Image captured by
a Nokia 6600 camera phone

G channel

The resultant image of
adaptive thresholding

B channel

Fig. 6.14   Effect of the adaptive thresholding (left) and the three thresholded colour channels (right). For a colour version, please see Plate 14.

However, the thresholded image also reveals that most cells originally in magenta appear to be in red, including their corresponding colour reference cell. Also, a few cells that were originally cyan have been replaced with blue. It is presumed that this was caused by inaccurate colour conversion between the different colour spaces and by the lossy JPEG file format, which preserves the smooth variation of tone and colour but removes information in the high-frequency components (e.g. edge information). That is, the image to be thresholded has already lost the determinants that make magenta 'magenta' and cyan 'cyan'.

The RGB values of red are (1, 0, 0), whereas those of magenta are (1, 0, 1). The difference between red and magenta is made by the blue component in the RGB value. Suppose that the threshold value 0.5 is selected to determine whether the blue component is 1 or 0. When the value of the blue component is 0.51, it will be normalised to 1, determining the colour 'magenta', whereas if the value is 0.49, it will be normalised to 0 determining the colour 'red'. When the blue value has been reduced to less than the threshold, owing to the image processing that has taken place, adaptive thresholding determines the colour accordingly. That is, although this adaptive thresholding is capable of removing the lighting effect even when the target image is not evenly illuminated, it cannot recover colour information that has already been lost.

The eight colours at the vertices of the RGB colour cube, namely, red, green, blue, cyan, magenta, yellow, black and white are usually considered to be the colours furthest apart from one another. However, our experimental results indicated that the colour values of red and magenta, and those of blue and cyan, in the reconstructed image in fact often become very close, which prevents the reading software from distinguishing them robustly. So, in order to operate a colour 2D barcode system robustly, we need to select colours that can preserve the maximum distance from their neighbouring colours throughout the image processing, regardless of the colour spaces through which a barcode symbol goes and regardless of the file format applied to store the image.

Colour selection scheme to enhance clear colour separation

We have discussed the three factors that have a detrimental effect on the robustness in reading colour 2D barcodes: inaccurate colour conversion across two or more colour spaces the file format and its corresponding compression and decompression applied to store the image data; and the lighting effect, which can change colour values.

Two possible solutions are the use of reference colours and of adaptive thresholding to remove the detrimental lighting effects on the image. Both solutions have been found effective to some degree but are not truly satisfactory.

This may be the reason why most researchers have eventually opted to develop black and white 2D barcode, even though they know the potential advantages of using colour symbols to encode data, e.g. an enhancement in symbol design and an improvement in data capacity without increasing the physical symbol size. In a black and white format, the separation between black and white colours in terms of colour space distance is greatest and only these two colours need to be discriminated. This ensures reading robustness even though most black and white 2D barcode symbols go through the same image capture process as coloured 2D barcode symbols.

Initially, we made the assumption that the eight colours red, green, blue, cyan, magenta, yellow, black and white were the maximum distance apart in the RGB colour space and thus it is easier to distinguish them than the colours in any different set. Consequently, they were the candidates for the colour symbols of our prototype 2D barcode and were used to examine the efficiency of adaptive thresholding. However, as mentioned earlier, examination revealed that the values of particular colours such as red and magenta and blue and cyan often become very close after going through the conversion between colour spaces and lossy compression and decompression, even when the lighting effect is considerably reduced.

This led to the conclusion that we need to select a set of colours that can preserve maximum distances between neighbours across all the colour spaces involved in the barcode operating process and, furthermore, through lossy compression and decompression. The question then is how to select these colours. It was found that not all the eight colours that are furthest apart in the RGB space remain so after colour conversion and data compression and decompression. So, in order to preserve the initial distances, some colours need to be removed, limiting the number of colour symbols that can be used for robustly encoding and decoding data and resulting in less data density for a given space.

Thus the outcome of our experiment suggests that a slight inaccuracy or trivial information loss in 3D space may result in a considerable change in colour values compared with that in 2D space, when colours are converted from one colour space to another and/or the colour data are compressed or decompressed. This may explain why the distances between some colours have decreased after these operations. As an attempted solution, we selected colours from a 2D space, i.e. a plane of the RGB colour cube, then made colour conversions between the planes of each colour space instead of a 3D to 3D conversion. In order to choose the best plane for our purpose, we examined the susceptibility of each of the eight colours at the vertices of the RGB colour cube to the colour conversion and data compression and decompression effect. This examination was conducted as follows.

(i) Eight 5.0×5.0 cm^2 squares, each in one of the eight colours at the vertices of the RGB colour cube (we will refer to them as colour squares hereafter), were produced and printed.

(ii) The printed sampled colour squares were captured using two camera phones, a Nokia 6110 and a Nokia 6300.

(iii) The sampled colour squares displayed on a computer screen were captured using a camera phone and the outcomes were compared with the printed samples. Since the results produced by the two camera phones were quite close in the second step, only one camera phone (i.e. the Nokia 6110) was used for the rest of the investigation.

(iv) The R, G and B components of the reconstructed image of each sampled colour were examined by computing their minimum, maximum and average values.

(v) Their respective thresholds are set to the mean value of the average of the R, G and B components of the reconstructed images whose original value equals 0 and the average of the R, G and B components whose original value equals 1. Although the original RGB colour values were defined within the range [0, 1], subsequent computation was performed for the RGB colour range, [0, 255].

(vi) An investigation was made to discover which sampled colours were most likely to return errors when each R, G and B component of their reconstructed image was normalised according to the threshold set in step (v).

The room used for the experiment was lit by fluorescent white lights.

The experimental results (see Figures H.1 and H.2 in Appendix H) reveal the following.

(i) Among the eight colours, red (R), green (G), blue (B), cyan (C), magenta (M), yellow (Y), black (K) and white (W), the colours B, C and M are more susceptible to colour-conversion and data compression and decompression effects than the other colours; thus, their use should be avoided. In contrast, K is the best colour to be used as its colour values are consistent, demonstrating its robustness.

(ii) The RGB values of M are quite similar to those of R.

(iii) The RGB values of C are very close to those of both B and G.

(iv) When comparing the three primary colours, red is least susceptible and blue is most susceptible to the effects of colour conversion and also of compression and decompression.

(v) There are six different ways to cut the RGB colour cube to maximise the space of each plane (i.e. six diagonal cuts). These are planes with corners BMYG, KBWY, KGWM, KRWC, RMCG and RYCB. When comparing these different cuts, on the one hand the colour values of the plane KRWC provided the points furthest from each other and less clustered than those of the other planes. On the other hand, the plane RMCG produced the worst results.

(vi) The RGB values of the reconstructed images of the printed samples were between 0 and 130, whereas those of the samples displayed on the computer screen were between 60 and 180. This confirms the need for colour reference cells.

As a result, we selected the plane KRWC, and nine points on this plane whose values are equal to the values of the colour symbols to encode data. These points are (0, 0, 0), (0.5, 0, 0), (1, 0, 0), (0, 0.5, 0.5), (0.5, 0.5, 0.5), (1, 0.5, 0.5), (0, 1, 1), (0.5, 1, 1) and (1, 1, 1) in the normalised RGB colour range [0, 1], representing black, brown, red, dark green, grey, tan, cyan, sky blue and white, respectively. Figure 6.15 illustrates these nine symbols plus yellow (see the next subsection) represented in the RGB colour space and the same 10 symbols represented in the YCbCr colour space.

As explained previously, values in the YCbCr colour space outside the RGB colour space are considered to be invalid and are processed so as to generate valid RGB values during the colour conversion process. The 10 colours of YCbCr space shown in Figure 6.15 lie on the boundary of the valid RGB values, at the maximum distances apart within the valid limit. This demonstrates visually that these 10 colours are furthest apart across more than one colour space, i.e. the RGB and YCbCr colour spaces in this case. The clear separation of the colour symbols enabled by this novel colour-selection scheme helps to improve the robustness in reading colour 2D barcodes even when resource-limited camera phones are used as image processing devices.

Additional colour symbol and its brightness

As mentioned above, in addition to the nine colours selected from the single plane KRWC, yellow was also selected, resulting in 10 colour symbols available for encoding data. Throughout the colour selection process, the colour values of yellow were always accurately retrieved, demonstrating its robustness to any operations applied.

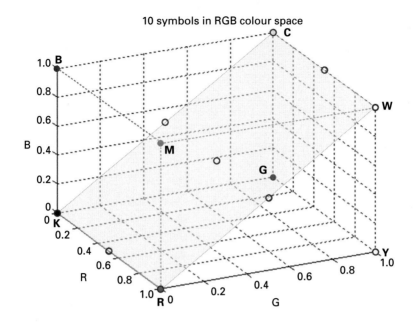

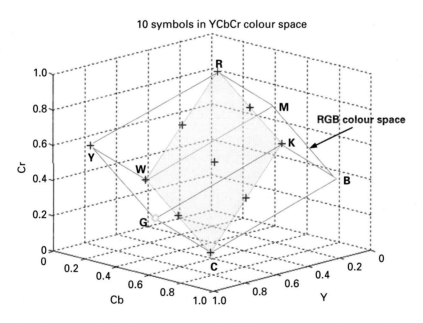

Fig. 6.15   The 10 colour symbols (circles) represented in RGB colour space and the same 10 colours represented in YCbCr colour space (crosses).

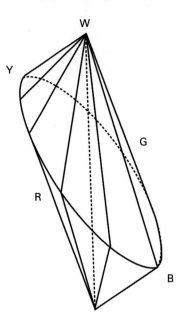

Fig. 6.16   Representation of hue–chroma–brightness space by Kirschman (adapted from [70]).

A possible reason for this is the distinctive brightness of yellow as compared to other colours. In some colour spaces where a sphere or symmetrical cone is used to represent colours, all the pure colours with identical brightness values are placed on one plane. According to [70], however, different hues (pure colours) reach their maximum chrominance at different tonal levels. Consequently, to represent the actual brightness of each pure colour along the axis that indicates brightness (or value or intensity), the colour wheel should be tilted through space in such a way that yellow occupies a high position opposite light grey and blue occupies a low position opposite dark grey [70]. Figure 6.16 shows a skewed double cone, a solution first suggested by Kirschman that represents the actual brightness of pure colours [70]. It clearly shows that yellow is nearly as bright as white, whose brightness is 100%.

6.2.3   Code design

MMCC™ code structure

The MMCC™ barcode consists of two major code areas: a *header information area* encoded into the 1D barcode symbol and a *data area* encoded into a colour 2D barcode symbol. The header information area can include MMCC™ barcode-encoding information such as:

(i) the symbol version, defining the format information such as the number of data cells and the position of the colour reference area;

(ii) the level of error correction;

Table 6.2. Code mapping table

Decimal	Colour	Colour value
0	yellow (Y)	(1, 1, 0)
1	white (W)	(1, 1, 1)
2	sky blue (S)	(0.5, 1, 1)
3	cyan (C)	(0, 1, 1)
4	tan (T)	(1, 0.5, 0.5)
5	grey (G)	(0.5, 0.5, 0.5)
6	dark green (D)	(0, 0.5, 0.5)
7	red (R)	(1, 0, 0)
8	brown (B)	(0.5, 0, 0)
9	black (K)	(0, 0, 0)

(iii) the security settings; and

(iv) an indicator stating whether a structured append, has been used.

Depending on the amount of the header information, the 1D barcode can be replaced with a black and white 2D matrix barcode.

The data area includes two different data types; the original data to be encoded and the data for error detection and correction. The MMCC™ barcode uses a Reed–Solomon code for its error detection and correction. The reading software first retrieves the header information from the 1D barcode, and this is followed by retrieval of the colour values of each colour reference cell. Once the colour information is acquired, it is used to decode the data in the data area.

The integration of the header information barcode (either 1D barcode or 2D barcode) and of the colour reference area into the finder pattern is one of the novel features of the MMCC™ barcode.

Colour data symbols

The proposed colour selection scheme allows us to use up to 10 colour symbols to encode and decode data robustly. This in turn allows us to encode decimal data, each colour representing an integer from 0 to 9, thus composing a data cell. Table 6.2 is an example of the data-to-colour mapping method, according to which the decimal data are converted to the corresponding colour symbols. The sets of colour values are presented in the normalised range [0, 1].

6.3   MMCC™ barcode encoding algorithm

The MMCC™ barcode encoding algorithm is based on the code design. Five main steps are involved in generating an MMCC™ barcode symbol. They are:

(i) compressing the raw data to be encoded;

(ii) error-protecting the compressed data;

   (iii) mapping the interleaved data onto colour symbols;
   (iv) creating a header information barcode; and
   (v) generating an MMCC™ barcode symbol.

These steps will now be discussed in turn.

Data compression

Either raw data or compressed data can be encoded into an MMCC™ barcode symbol. In general, however, data compression is performed prior to the encoding of data as this can considerably increase the data capacity of the MMCC™ barcode symbol for a given space. Different data compression algorithms can be used depending on the type of raw data.

Error-protecting the data to improve the robustness of MMCC™ barcode

As explained in Chapter 5, the performance of Reed–Solomon codes can be significantly improved by the addition of interleaving. Hence, both Reed–Solomon coding and a technique such as pseudo-random interleaving are used for error protection. The error correction level can be selected according to the symbol's requirement. However, an error correction rate of approximately 22% is the standard.

Once the original data have been compressed, Reed–Solomon codes are computed and added to the compressed data. Eight bits are used to encode each symbol, resulting in a data range between 0 and 255. Prior to applying the pseudo-random interleaving operation, these data are converted to decimal integers and padded with '0's as required. This zero-padding is performed when the amount of data is not sufficient for encoding into a particular version of the MMCC™ symbol. For example, when the number of integers (i.e. the data to be encoded) is less than 900 for an MMCC™ barcode symbol with 30 × 30 data cells, zero-padding is performed. After all necessary operations have been completed, the entire data set is interleaved.

Mapping from decimal data to colour symbols

The interleaved data are mapped to colour symbols according to the code conversion algorithm. Table 6.2 presents an example of the MMCC™ code mapping method.

Encoding header information

A header information barcode can be either a 1D barcode or a 2D matrix barcode. Encoding of the symbol version is essential for correct decoding of an MMCC™ barcode since this defines the number of data cells and the position of the colour reference area. Additional information such as the error correction level, the security settings and a structured append indicator can also be encoded.

Generating an MMCC™ symbol

The MMCC™ symbol can be either printed out or displayed on a digital screen. Figure 6.17 illustrates an example of the resultant MMCC™ barcode symbol with 30 × 30 data cells.

Plate 1   ColorCode design variation (adapted from [41]).

Plate 2   Examples of mixed ColorCode (adapted from [41]).

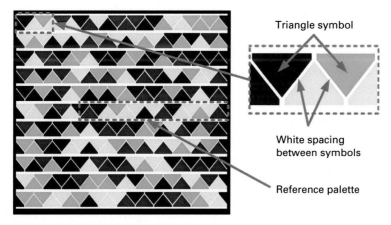

Plate 3   Symbol structure of HCCB.

Plate 4   Paper Memory (PM) Code symbol and its layered structure.

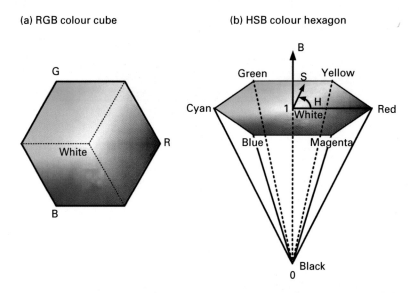

Plate 5   (a) Red, green, blue (RGB) colour cube and (b) hue, saturation, brightness (HSB) colour hexagon cone, defined by red, yellow, green, cyan, blue and magenta axes radiating from the origin (black).

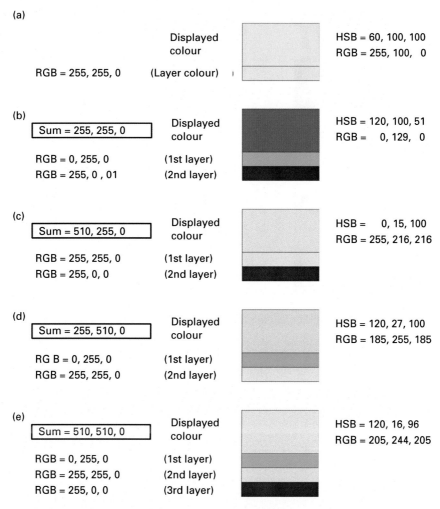

(a)

Displayed colour

HSB = 60, 100, 100
RGB = 255, 100, 0

RGB = 255, 255, 0   (Layer colour)

(b)

Sum = 255, 255, 0

Displayed colour

HSB = 120, 100, 51
RGB =   0, 129, 0

RGB = 0, 255, 0   (1st layer)
RGB = 255, 0 , 01   (2nd layer)

(c)

Sum = 510, 255, 0

Displayed colour

HSB =   0, 15, 100
RGB = 255, 216, 216

RGB = 255, 255, 0   (1st layer)
RGB = 255, 0, 0   (2nd layer)

(d)

Sum = 255, 510, 0

Displayed colour

HSB = 120, 27, 100
RGB = 185, 255, 185

RG B = 0, 255, 0   (1st layer)
RGB = 255, 255, 0   (2nd layer)

(e)

Sum = 510, 510, 0

Displayed colour

HSB = 120, 16, 96
RGB = 205, 244, 205

RGB = 0, 255, 0   (1st layer)
RGB = 255, 255, 0   (2nd layer)
RGB = 255, 0, 0   (3rd layer)

Plate 6   PM Code colour conversion (adapted from [59]); see also Figure 4.16.

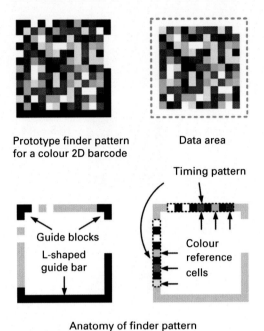

Prototype finder pattern
for a colour 2D barcode

Data area

Timing pattern

Guide blocks

L-shaped
guide bar

Colour
reference
cells

Anatomy of finder pattern

Plate 7   Structure of a prototype finder pattern.

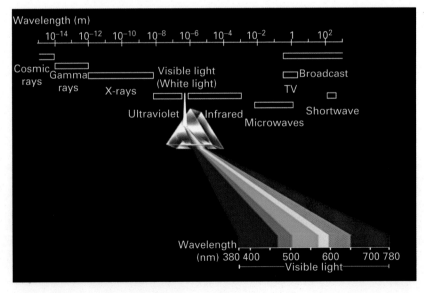

Plate 8   Colour spectrum (adapted from [65]).

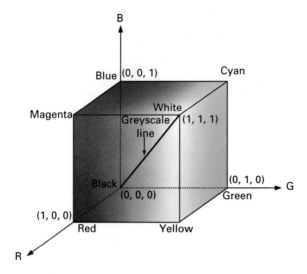

Plate 9   RGB colour cube.

Additive colour mixture of
RGB primary colours of *light*

Subtractive colour mixture of
CMY primary colours of *pigment*

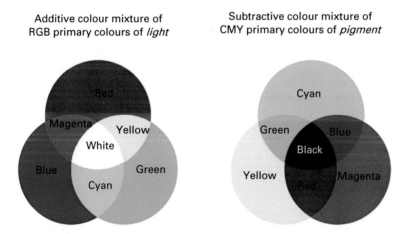

Plate 10   RGB additive colour space and CMY subtractive colour space.

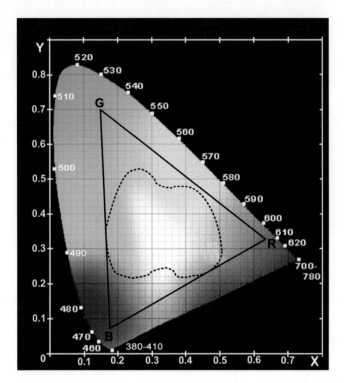

Plate 11   Colour gamuts of colour monitors (solid black line) and colour printing devices (broken black line), for comparison. The horseshoe of colours corresponds to the visible range of the electromagnetic spectrum (adapted from [64]).

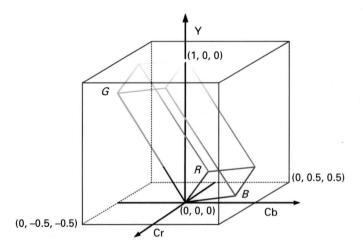

Plate 12   Relation between RGB and YCbCr colour space (adapted from [67]).

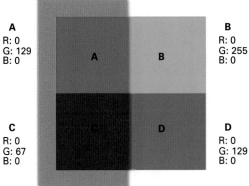

Plate 13   Effect of lighting on colour values.

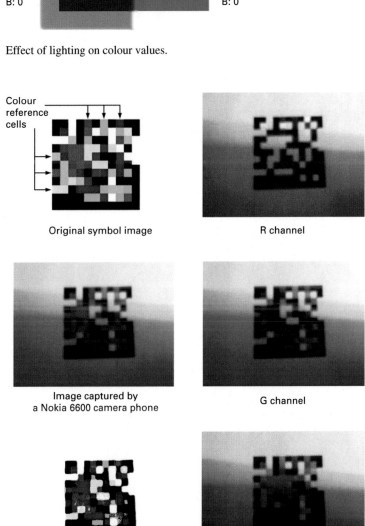

Plate 14   Effect of the adaptive thresholding (left) and the three thresholded colour channels (right).

Plate 15   Example of a complete MMCC™ symbol image with 30 × 30 data cells.

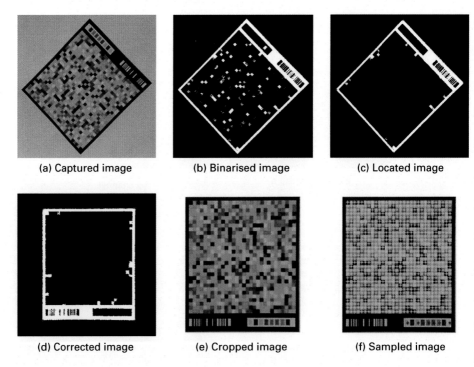

(a) Captured image   (b) Binarised image   (c) Located image

(d) Corrected image   (e) Cropped image   (f) Sampled image

Plate 16   Decoding process of the MMCC™ barcode.

Number of cells	30 x 30 cells
Symbol size	5.5 x 5.0 cm^2
Error correction rate	Approximately 22%
Data type	Numeric data
Original data size	11.7 KB

Plate 17   Sample MMCC™ symbol for the second experiment.

Original image (Sample A1)

QVGA	
PSNR	26.4746
MSE	146.4267

VGA	
PSNR	26.0691
MSE	160.7555

1.3 megapixels	
PSNR	26.1339
MSE	158.3775

Plate 18 The resulting images after orientation and distortion correction of the captured images. The images produced by the VGA and 1.3 megapixel camera settings are reduced to the image size of QVGA. The fidelity of the reconstructed images is measured by the PSNR and MSE.

(a) Torn symbols

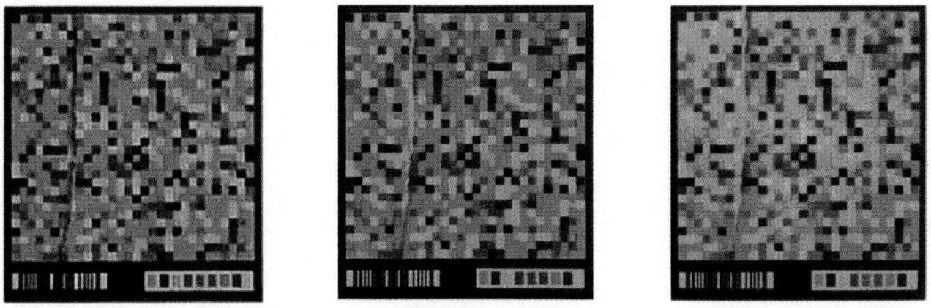

QVGA (320 x 240)   VGA (640 x 480)   1.3 megapixels (1280 x 960)

(b) Smudged symbols

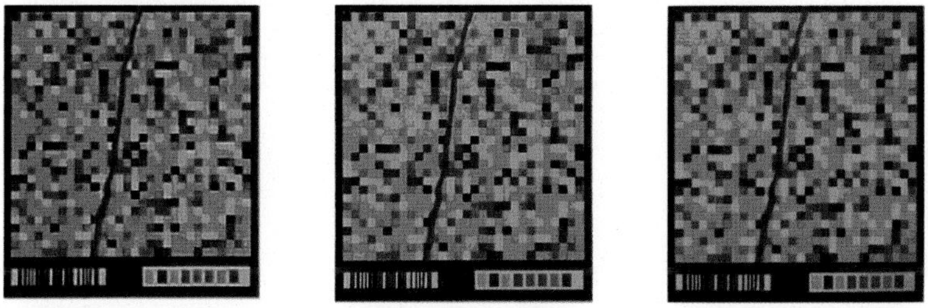

QVGA (320 x 240)   VGA (640 x 480)   1.3 megapixels (1280 x 960)

(c) Symbols having three voids

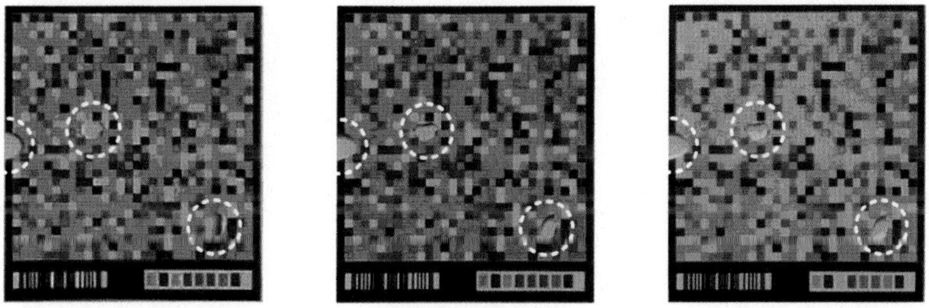

QVGA (320 x 240)   VGA (640 x 480)   1.3 megapixels (1280 x 960)

(d) Symbols having two voids captured under the effect of incandescent light

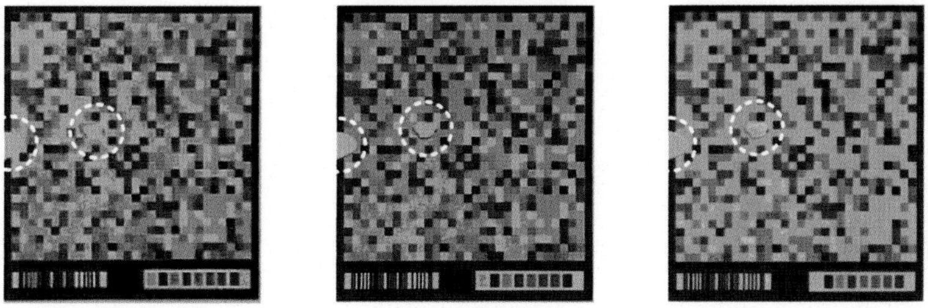

QVGA (320 x 240)   VGA (640 x 480)   1.3 megapixels (1280 x 960)

Plate 19   Legible MMCC™ symbols that have defects.

(a) Under the effect of incandescent light

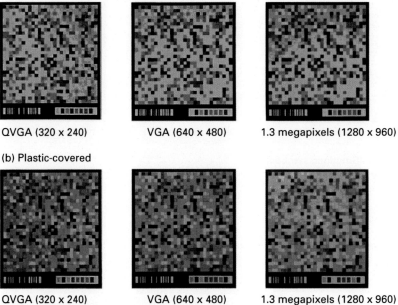

QVGA (320 x 240)   VGA (640 x 480)   1.3 megapixels (1280 x 960)

(b) Plastic-covered

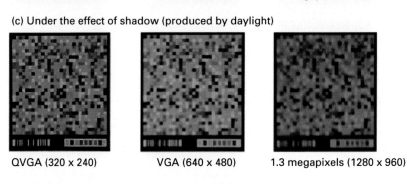

QVGA (320 x 240)   VGA (640 x 480)   1.3 megapixels (1280 x 960)

(c) Under the effect of shadow (produced by daylight)

QVGA (320 x 240)   VGA (640 x 480)   1.3 megapixels (1280 x 960)

(d) Under the effect of shadow (produced by incandescent light at night)

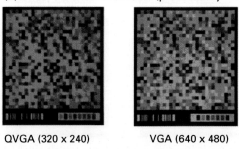

QVGA (320 x 240)   VGA (640 x 480)

Plate 20   Legible MMCC™ symbols under illumination effects.

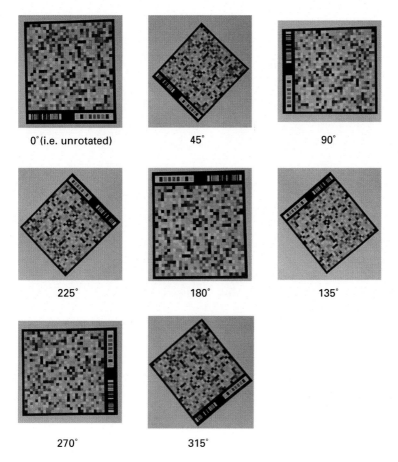

Plate 21   Legible MMCC™ symbols in different orientations. Rotation angles are measured clockwise from the upright position.

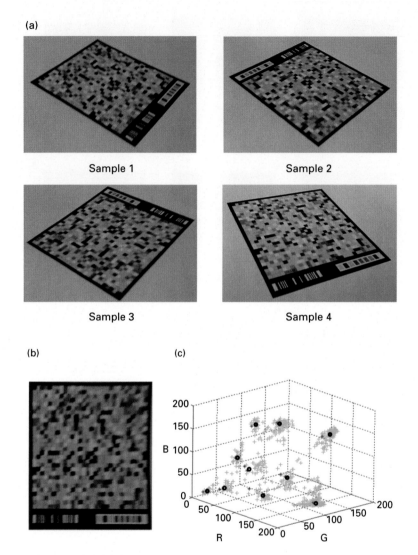

(a)

Sample 1

Sample 2

Sample 3

Sample 4

(b)

(c)

Plate 22 (a) Legible MMCC™ symbols with different distortions. (b) Sample 1 after the distortion correction was performed and (c) its colour value distribution.

(a) Crumpled symbols

QVGA (320 x 240) VGA (640 x 480)

(b) Shadow cast on the MMCC™ symbol by blocking an incandescent light

1.3 megapixels (1280 x 960)

(c) Shadow partially cast on the MMCC™ symbol

QVGA (320 x 240) VGA (640 x 480) 1.3 megapixels (1280 x 960)

(d) Distorted symbol having curved edges

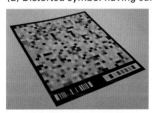

Plate 23   Illegible MMCC™ symbols.

(a) Illegible crumpled symbol

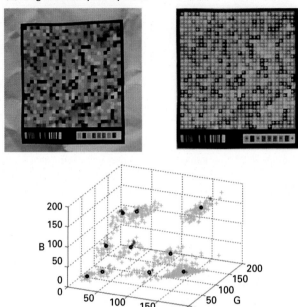

(b) Legible crumpled symbols

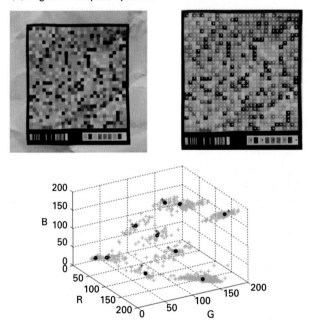

Plate 24   Examples of (a) an illegible and (b) a legible crumpled MMCC™ symbol. The right-hand figures show sampled versions of the images.

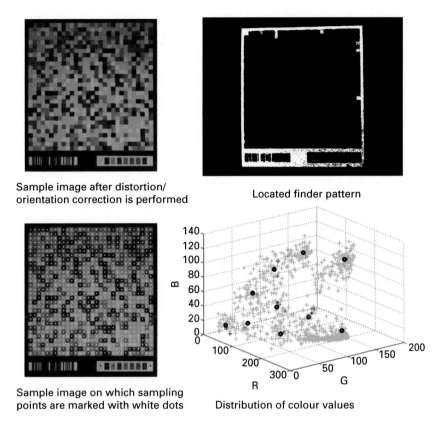

Sample image after distortion/
orientation correction is performed

Located finder pattern

Sample image on which sampling
points are marked with white dots

Distribution of colour values

Plate 25   Illegible MMCC™ symbol entirely under the effect of a shadow cast by an incandescent light.

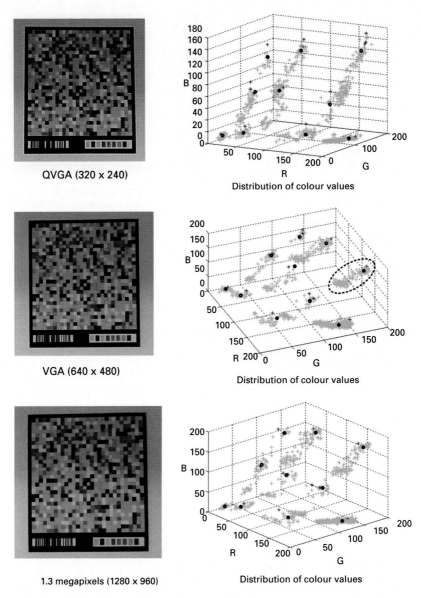

QVGA (320 x 240)

Distribution of colour values

VGA (640 x 480)

Distribution of colour values

1.3 megapixels (1280 x 960)

Distribution of colour values

Plate 26   Illegible MMCC™ symbols partially in shadow.

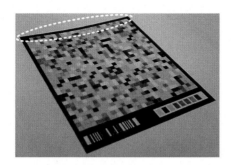

Sample image captured by
VGA camera resolution

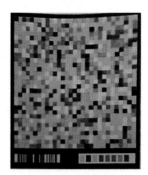

Sample image after distortion/
orientation correction

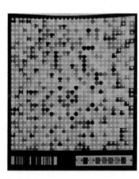

Sample image on which sampling
points are marked with white dots

Distribution of colour values

Plate 27   Illegible distorted MMCC™ symbol with a curved edge.

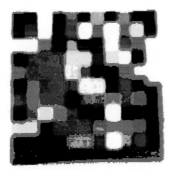

Plate 28   An image under the every-other-line effect due to the zigzag scanning of an adaptive
thresholding method.

Fig. 6.17   Example of a complete MMCC™ symbol image with 30 × 30 data cells. For a colour version, please see Plate 15.

6.4   MMCC™ barcode decoding algorithm

The MMCC™ barcode decoding algorithm consists of two major parts: the MMCC™ symbol-recognition algorithm and the MMCC™ decoding algorithm. The former mainly defines the procedure for locating the target symbol and correcting its orientation and distortion, whereas the latter is the procedure for decoding the colour symbols accurately.

6.4.1   MMCC™ barcode symbol-recognition algorithm

Figure 6.18 illustrates the decoding process for the MMCC™ barcode. This prototype colour 2D barcode is designed for use with mobile devices, especially, camera phones. Hence the recognition algorithm was developed for images captured by camera phones. Once an MMCC™ barcode symbol is captured (see Figure 6.18(a)), the symbol-recognition operation takes place in the following steps.

  (i) The input image is thresholded and binarised (Figure 6.18(b)).
 (ii) The largest continuous region within the input image is located as a candidate for the finder pattern of the MMCC™ barcode symbol (Figure 6.18(c)).
(iii) The orientation of the candidate finder pattern is corrected (Figure 6.18(d)).
(iv) A projective transformation is performed to correct symbol distortion.
 (v) The MMCC™ barcode symbol area is separated from the background image by cropping (Figure 6.18(e)) and the value of each data cell is sampled.

These steps are considered below in turn.

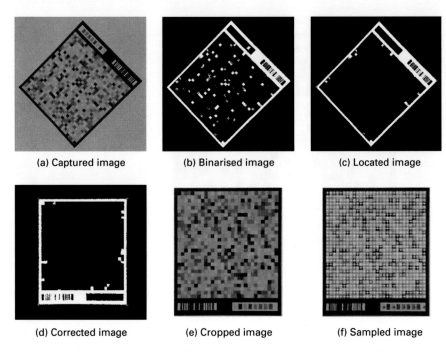

(a) Captured image	(b) Binarised image	(c) Located image
(d) Corrected image	(e) Cropped image	(f) Sampled image

Fig. 6.18   Decoding process of the MMCC™ barcode. For a colour version, please see Plate 16.

Thresholding the captured image

Thresholding is performed to convert the reconstructed colour image to a binarised image, which facilitates the detection and correction of the target barcode symbol. This process removes the effect of colour, including the colours in the reference cells, leaving a skeleton of the finder pattern and any background components having similar colour values to the finder pattern. In our work, thresholding was conducted in two different ways, using the modified adaptive thresholding method [17] explained earlier and using the threshold obtained by experiment.

The former scheme showed itself as an effective method to remove any uneven lighting effect from the input image. However, as was pointed out in [17] adaptive thresholding is very time-consuming even when floating point operations are replaced with shifted-integer operations to reduce the operation time. Furthermore, often the resulting image introduced more noise than the alternative scheme. In the proposed recognition algorithm, binarisation is performed to locate a candidate finder pattern properly and to correct its orientation and/or distortion. In our observation, there was no significant difference in the fidelity of images thresholded by adaptive thresholding and by our alternative method (see Figure 6.19). Moreover, the alternative method can save considerable computational power. Consequently, we adopted this alternative thresholding scheme.

Searching for the finder pattern

The finder pattern is the solid bounding border, in which a separate barcode, for the header information, and the colour reference cells are embedded. The reading software

Binarised images using two different thresholding methods

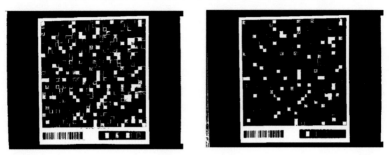

Fig. 6.19   Binarised images using adaptive thresholding (left) and using a set of threshold values obtained by experiment (right).

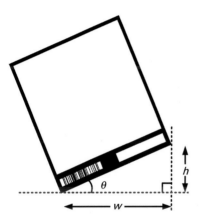

Fig. 6.20   Orientation of the MMCC™ symbol.

first searches for the largest of all the regions found within the captured image. Once this region is located, the coordinates of its four corners are calculated and these are used to measure the side lengths of the rectangular finder pattern. These lengths and the coordinates of the four corners enable the orientation of the located finder pattern to be calculated.

Correcting the orientation of the candidate finder pattern

The tangent of an angle in a right-angled triangle is the ratio of the length of the side opposite to the angle and the length of the adjacent side. The bottom part of the finder pattern, where the header barcode and the colour reference cells are embedded, can be found by computing the width of each side of the bounding border. The widest side is the bottom. The length h of the opposite side and the length w of the adjacent side can be obtained from the coordinates of both ends of the bottom (see Figure 6.20). Hence, the orientation angle θ between the x-axis of the image and the base of the bottom part can be calculated as

$$\theta = \arctan\left(\frac{h}{w}\right).$$

Performing a projective transformation to correct for symbol distortion

A projective transformation, also known as a homogeneous transformation or projective mapping [72], is a quadrilateral-to-quadrilateral mapping between the code plane and the image plane. It can be performed once the four corresponding points are located. The centroids of the cells at the four corners of the finder pattern are used for the calculation.

The general form of a projective transformation is a rational linear mapping [72] in which the code coordinates (u, v) are mapped to the image coordinates (x, y):

$$x = \frac{au + bv + c}{gu + hv + i}, \qquad y = \frac{du + ev + f}{gu + hv + i},$$

where we can assume $i = 1$ without loss of generality.

The eight unknown coefficients a to h, which denote the external parameters (e.g. position and direction) and the internal parameters (e.g. the focal length) of the camera [73], can be obtained from the four-point correspondences between quadrilateral corners:

$$
\begin{pmatrix} x_0 \\ x_1 \\ x_2 \\ x_3 \\ y_0 \\ y_1 \\ y_2 \\ y_3 \end{pmatrix} =
\begin{pmatrix}
u_0 & v_0 & 1 & 0 & 0 & 0 & -x_0u_0 & -x_0v_0 \\
u_1 & v_1 & 1 & 0 & 0 & 0 & -x_1u_1 & -x_1v_1 \\
u_2 & v_2 & 1 & 0 & 0 & 0 & -x_2u_2 & -x_2v_2 \\
u_3 & v_3 & 1 & 0 & 0 & 0 & -x_3u_3 & -x_3v_3 \\
0 & 0 & 0 & u_0 & v_0 & 1 & -y_0u_0 & -y_0v_0 \\
0 & 0 & 0 & u_1 & v_1 & 1 & -y_1u_1 & -y_1v_1 \\
0 & 0 & 0 & u_2 & v_2 & 1 & -y_2u_2 & -y_2v_2 \\
0 & 0 & 0 & u_3 & v_3 & 1 & -y_3u_3 & -y_3v_3
\end{pmatrix}
\begin{pmatrix} a \\ b \\ c \\ d \\ e \\ f \\ g \\ h \end{pmatrix}. \tag{6.3}
$$

Separating the MMCC™ symbol area from the background image

Once all the necessary corrections are performed, the next step is decoding the barcode. Prior to this, the image is cropped leaving only the 2D barcode symbol, which improves the subsequent performance. This operation is applied to the coloured image, not the binarised one.

6.4.2 MMCC™ barcode decoding algorithm

In order to maximise data capacity within a given space, the data area of the proposed prototype 2D barcode does not include the format information. Prior to reading the data area of the symbol, therefore, the header information such as the number of cells and the error correction level must be retrieved from the barcode embedded into the finder pattern. The decoding process of an MMCC™ barcode takes place in the following steps.

(i) The header information is retrieved.

(ii) The colour values of the colour reference cells in their known positions are sampled (Figure 6.18(f)).
(iii) The colour value of the centroid of each data cell is read.
(iv) The colour of each data cell is determined.
(v) The colour coordinates are mapped to the corresponding data values.
(vi) Error detection and correction is performed.
(vii) The raw encoded data are retrieved.

Retrieving the header information

The header information can include optional information such as security settings and an indicator regarding the use of the structured append. Of these, the essential information is the symbol version, which defines the number of cells in the data area, and the position of the colour reference area. Together with the computed symbol size (i.e. its height and width), this information enables the calculation of the centroid of each data cell. The positions of the colour reference cells are fixed for a given symbol version.

Sampling the colour values of the colour reference cells

There are two sampling approaches to retrieve the colour values: *single-pixel sampling* and *plural-pixel sampling*. In the former approach, one simply samples the centroid of each cell and retrieves the set of colour values that belong to that single pixel. In the latter approach, several pixels within each cell are sampled and the averages of the R, G and B components of the sampled values are considered to represent the colour values of that cell.

If, for example, there happens to be an ink stain on the sampled pixel and the colour values are determined accordingly then referring the values of that single pixel may result in incorrect colour determination of the data cell. To avoid this possibility in regard to sampling the colour values of the reference cells, the following approach was adopted. The average is taken of the colour values of the nine sampled pixels in the 3×3 window whose centre is the centre pixel of the target cell. In order that a stained pixel does not change the whole colour value, any extremes are removed by computing the gaps in colour value between adjacent pixels before calculation of the average.

Reading the colour values of the centroid of each data cell

To read the centroid of each data cell, first the positions of the centroids must be computed.

As illustrated in Figure 6.21, the width of the solid bounding border is equivalent to the side of a data cell. The depth of the bottom part of the finder pattern is five times the side of a data cell. Hence, the height H_c and width W_c of each data cell can be computed from the height H_f and width W_f of the finder pattern and from the number of data cells:

$$H_c = \frac{H_f}{\text{number of cells} + 6}$$

and

$$W_c = \frac{W_f}{\text{number of cells} + 2}.$$

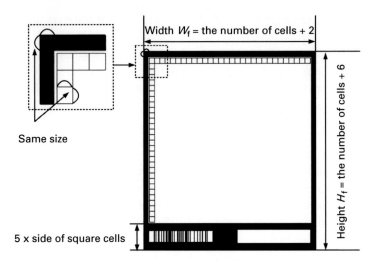

Width W_f = the number of cells + 2

Height H_f = the number of cells + 6

Same size

5 x side of square cells

Fig. 6.21　　Properties used for computing the position of the centroid of each data cell.

Although the data cells start out square, the cell height and width are calculated separately as the shape may be slightly changed after orientation and distortion correction. This results in accurate computation of the centroids (see Figure 6.18(f)).

In contrast with the sampling approach taken in the second step, a single-pixel sampling is applied to reading data cells, for two reasons. The first reason is that this saves computational power. Another reason is that it may reduce the likelihood of introducing unreliable colour values. As compared with the reference cells, the data cells are rather small. Furthermore, there are no separators between the data cells, which may lead to colour blending at the edges of adjacent data cells. The centroid of each cell is the furthest point from the blending effect and, as a result, the set of colour values of the centroid is considered to be the most reliable information.

Owing to possible computational errors (e.g. a cumulative error), the reading software may sample the colour value of a pixel slightly away from the real centroid. If, for example, the centre of the 3×3 window is off the correct point, some pixels around it will be even further away. This means that they are closer to the edge pixels and under the blending effect, increasing the chance of sampling unreliable colour values.

Determining the colour of each data cell

Once the sampling has been completed, the next step is determining the colour of each data cell. This can be achieved by comparing the set of sampled colour values with those in the colour reference cells. The distances between the sampled colour values and those of each reference cell in the 3D RGB colour space are first calculated. Then, the colour of the sampled cell is determined by selecting the reference colour whose colour values are closest to those of the data cell.

The use of colour reference cells is essential for robust colour barcode reading. However, it may not be perfectly reliable. It appears that the colour values of reference cells are often lower than those of data cells. A possible cause for this is the location of the

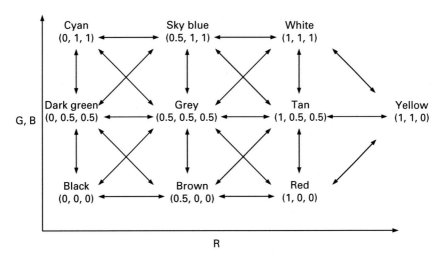

Fig. 6.22   Positions of the 10 colours in the 2D plane.

colour reference area. Although the reference cells are enclosed by a four-pixel-wide white separator, the reference area is embedded in the black finder pattern, which can darken the entire area.

  To solve this problem, we developed a colour sampling scheme to determine new reference points. They can be obtained in the following steps.

(i) The distance between each reference point and its neighbouring reference points is computed. The neighbouring colours of each reference colour are presented in Figure 6.22.

(ii) The minimum distance between each reference point and its neighbouring colour is found.

(iii) An imaginary circle whose radius is equivalent to 25% of the minimum distance from each reference point is drawn. The percentage is increased by 5% until the radius reaches 50% of the minimum distance.

(iv) The colour determination process is performed, making colour clusters for each colour. This is continued until either the number of colours assigned to each reference colour cluster reaches 3% of all the colour symbols or the radius of each circle reaches 50% of the minimum distance.

(v) The average of the colour values is calculated for each colour cluster, which results in a new reference point for the corresponding colour.

(vi) The new reference point is validated. The minimum distance previously obtained is compared with the distance between the new reference point and its closest neighbouring reference point. If the former is shorter then the new reference point is considered to be valid; otherwise it is considered invalid. If the new reference point is invalid, the previous reference point is used for the subsequent colour determination process.

Once the new reference points are acquired, the entire colour determination process is repeated.

Mapping the colour symbols onto the corresponding data value

The mapping between colour symbols and the corresponding data values can be done simply using the code conversion table (see Table 6.2). This operation is the inverse of the MMCC™ barcode encoding process.

Performing error detection and correction

In order to increase the robustness of the MMCC™ barcode, a combination of Reed–Solomon coding and interleaving is used on encoding, to protect the data. Thus, when decoding, to detect and correct errors the data must first be de-interleaved. The de-interleaved data are then corrected if errors are found.

Retrieving the raw encoded data

In general, the data are compressed before being encoded in an MMCC™ barcode symbol. Hence, the data are first decompressed and then, the original data are retrieved. Figures I.1 and I.2 in Appendix I present flowcharts that illustrate the method of encoding information into an MMCC™ barcode symbol and the method of retrieving the original information from an MMCC™ barcode symbol.

6.5   Summary

The goal of our research was to develop a colour database 2D barcode for camera phone use with the following advantageous features:

  (i) higher data capacity than any existing 2D barcode;
 (ii) tolerance of physical damage; and
(iii) robust and reliable operation for resource-limited mobile devices.

The solution proposed for achieving the first goal is the use of colour symbols to encode data. In order to improve significantly the data capacity or, more precisely, the data density of a barcode within a given fixed space, our challenge was to develop a colour 2D barcode that uses more colours than the number of colours previously used for camera phones.

Throughout the designing phase, including symbol design and code design, our focus was on how to improve the data capacity for a given space while preserving the reading robustness of a colour 2D barcode. The shape of the symbol and the data cell were determined as square, to make the most of the available space on a camera phone display.

A solid bounding border was selected as the finder pattern of the MMCC™ barcode to ensure its robustness even when partially damaged or when the ratio of the area it occupies and that of the entire image is reduced owing to, for example, the barcode symbol having been captured from a distance. The pursuit of ways to maximise the data capacity resulted in the novel feature of the finder pattern whereby the header information barcode and the reference area for colour calibration are included within the finder pattern border. In this way the reading robustness of barcode symbols is increased, since the black and white barcode in which the header information is encoded can be decoded more robustly than a coloured barcode.

The most challenging task throughout the entire barcode development was the selection of the colours to be used for encoding data. The more colours are used in a colour space, the greater the reduction in distance between neighbouring colours, which makes it difficult to distinguish all the colours accurately. To make it worse, there are three factors that have a detrimental effect on the robustness of reading colour 2D barcodes: inaccurate colour conversion between colour spaces; inadequacies of the file format and the corresponding compression and decompression that is applied to store the image data; and the lighting effect, which can change colour values.

One possible solution to overcome such problems is to use reference colours. However, this only works if the colour reference area and the data area are under the same lighting effect. If, for example, a barcode symbol is partially covered by a shadow then the colour values under the shadow may be significantly different from those outside the shadow, resulting in a failure to read the correct colour values. It appears that adaptive thresholding is capable of removing a great deal of such lighting effects. However, investigation revealed that the values of some colours that were originally different had become indistinguishable even after the lighting effect was satisfactorily removed. This indicated that some important colour information is lost through the operations involved in the barcode system (e.g. printing and image reconstruction). It also brought us to the view that colours at a maximum distance between each other in one colour space are not necessarily furthest apart in other colour spaces. This idea urged us to address a colour-selection scheme that ensures a maximum distance between colours across more than one colour space.

In a prior observation, some of the eight colours in the RGB colour space, i.e. red, green, blue, cyan, magenta, yellow, black and white, which are believed to be furthest apart, under colour conversion failed to preserve the distance between their neighbouring colours (e.g. red and magenta). Taking this as a clue, we selected colours from a plane of the RGB colour space, rather than using the RGB colour space itself. The experimental results showed positive outcomes, supporting our view that a plane-to-plane colour conversion can preserve the distance between colours more accurately and robustly than a space-to-space conversion.

The proposed novel colour selection scheme enabled the clear separation of 10 colours, providing 10 colour symbols for encoding data. Consequently, we developed encoding and decoding algorithms for this prototype colour 2D barcode, the MMCC™ barcode, based on the use of the 10 colour symbols selected. When developing the encoding and decoding algorithms, we adopted some techniques to increase the robustness of the MMCC™ barcode. Next, in Chapter 7, we will evaluate the effectiveness and robustness of these MMCC™ barcode encoding and decoding algorithms.

7 Evaluation of the prototype colour 2D barcode

As described in Chapter 6, a symbology known as the Mobile Multi-Colour Composite (MMCC™) 2D barcode has been developed by the authors. It is a novel colour 2D barcode that has the following attributes:

 (i)   a higher data density than any existing 2D barcode;
 (ii) tolerance to physical damage; and
 (iii) robust and reliable operation for resource-limited mobile devices.

The encoding and decoding algorithms of the MMCC™ barcode were described in Chapter 6. In the present chapter we evaluate the effectiveness and robustness of the MMCC™ barcode and its decoding algorithm.

7.1 Purpose of the experiments

In order to verify the effectiveness and robustness of the MMCC™ barcode and its decoding algorithm, we conducted experiments on the MMCC™ barcode symbols. Two sets of experiments were conducted to evaluate the MMCC™ symbology from different perspectives:

 (i) the overall performance of MMCC™;
 (ii) its effectiveness and robustness in different conditions and under a variety of scenarios.

To evaluate the performance of the MMCC™ barcode, in the first experiment, the first-read rate (FRR) of each sampled symbol was analysed, where

$$\text{FRR} = \frac{\text{Number of successful first reads}}{\text{Number of attempted first reads}}.$$

This metric allowed us to gauge the reading reliability of the MMCC™ quantitatively. The robustness of the barcode was examined by decoding sampled symbols in various conditions and under varying operating environments. A legible symbol in a certain condition or under a certain effect indicated that the MMCC™ symbology is tolerant of that type of condition or effect. Hence, the options 'legible' and 'illegible' were used as a metric to evaluate the robustness of the barcode in the second experiment.

7.2   Evaluation method and procedure

7.2.1   The first experiment – FRR analysis

Test samples

For the first experiment, 100 MMCC™ symbols were generated as samples (see Appendix J). Each MMCC™ barcode sample contains 165 bytes of alphanumeric characters in a matrix of 30 × 30 data cells, with a symbol size of 5.5 × 5.0 cm^2.

Equipment

This experiment used the following equipment:

(i) a Sanyo W33SA II camera phone;
(ii) a Hewlett–Packard (HP) Officejet 6310 all-in-one printer to produce each sample MMCC™ symbol;
(iii) MMCC™ encoder and decoder software written in MATLAB.

The camera resolution could be set to quarter video graphic array (QVGA, i.e. 320 × 240 pixels), VGA (i.e. 640 × 480 pixels) or 1.3 megapixels (i.e. 1280 × 960 pixels).

Experiment settings

The room used for the experiment was lit by daylight and no additional lighting was used. The sampled symbols were placed in the centre of a white desk and captured by the camera phone from above (see Figure 7.1).

Procedure

The experiment was conducted according to the following procedure.

(i) One hundred sample MMCC™ barcode symbols were generated using the encoding software.
(ii) The sample symbols were printed out using the inkjet printer.

Fig. 7.1   First experiment settings.

(iii) The sample MMCC™ symbols were captured using the camera phone, since the MMCC™ symbology is designed for camera phone use. Each sample symbol was captured in three different camera resolution settings to address the effect of camera resolution on the robustness and accuracy of symbol reading.

A non-skilled user performed the entire symbol-capture process. The user first performed several trials to get used to capturing each sample MMCC™ symbol in such a way that the symbol occupied approximately 90% of the phone screen. A certain amount of tilt and rotation was expected in the captured symbol images.

(iv) The captured symbols were decoded using the MMCC™ decoding software. The captured symbol images in the VGA and 1.3 megapixel camera settings were decoded in two ways, using the original image reading and using a subsampled image reading. In the subsampling operation, the size of an image is reduced to half of its original both horizontally and vertically, so that the result has one quarter of the original area (see Appendix K). Thus, after subsampling the VGA and 1.3 megapixel images were converted to QVGA and VGA images, respectively.

7.2.2   The second experiment – analysis of the reading robustness

Test samples

The second experiment used a 30 × 30 MMCC™ symbol to investigate the robustness of the MMCC™ 2D barcode. The details of the sampled symbol are presented in Figure 7.2. Defective symbols were created by tearing, making holes in, or drawing a line across, copies of the symbol.

Equipment

The following equipment was used in the second experiment:

(i) a Sanyo W33SA II camera phone;
(ii) a Hewlett–Packard (HP) Officejet 6310 all-in-one printer to produce each sample MMCC™ symbol;
(iii) MMCC™ encoder and decoder software written in MATLAB;
(iv) a 75 W 240 V incandescent lamp;
(v) a plastic sheet.

In addition to the equipment used in the first experiment, an incandescent lamp was used in the second experiment to capture the sampled symbol under different illumination conditions. The effect of a plastic cover over the symbol on the fidelity of the captured symbol and symbol reading was also observed.

Experiment setting

As before the room used for the experiment was lit by daylight and no additional lighting was used unless indicated otherwise. The sample symbols were captured in such a way as to reproduce a wide spectrum of potential image appearances, including defects and various illumination effects, orientations and distortions. The illumination effects were created in a room lit by an incandescent light.

Number of cells	30 x 30 cells
Symbol size	5.5 x 5.0 cm²
Error correction rate	Approximately 22%
Data type	Numeric data
Original data size	11.7 KB

Fig. 7.2   Sample MMCC™ symbol for the second experiment. For a colour version, please see Plate 17.

Procedure

The experiment was conducted according to the following procedure.

(i) The original data were compressed and encoded into a 30×30 MMCC™ symbol using the encoding software.

(ii) Copies of the sample symbol were printed out at a symbol size of 5.5×5.0 cm² using the inkjet printer.

(iii) Different types of physical defect were created on the printed symbols.

(iv) The sample MMCC™ symbols were captured using the camera phone.

The camera resolution was set to QVGA (i.e. 320×240 pixels), VGA (i.e 640×480 pixels) or 1.3 megapixels (i.e. 1280×960 pixels). As before, each sample symbol was captured in the three different camera settings to address the effect of camera resolution on the robustness and accuracy of symbol reading.

The user who performed the symbol-capture process in the first experiment also performed the symbol-capture process in the second experiment. Each sampled MMCC™ symbol was captured in the same way as in the first experiment, i.e. so that it occupied approximately 90% of the phone screen. Again, a certain amount of tilt and rotation was expected in the captured symbol images.

In order to provide varying lighting conditions, an incandescent light was used.

(v) The captured symbols were decoded using the MMCC™ decoding software.

As in the first experiment, the captured symbol images at the VGA and 1.3 megapixel camera settings were decoded in two ways, using the original image reading and using the subsampled image reading.

7.3   Experimental results and observations

7.3.1   The results of the first experiment

The first-read rates (FRRs) of the MMCC™ were 100% in all three different camera resolution settings. The FRRs of both subsampled images (i.e. the images converted from 1.3 megapixels to VGA and those from VGA to QVGA) also achieved 100%.

It is often believed that cameras with higher resolution perform better in reading barcode symbols. However, as we observed in our earlier studies [37], [74], higher camera resolution was not very important for reading MMCC™ barcode symbols. These results support our earlier claims and debunk the myth that expensive, high-resolution, cameras are needed to read colour 2D barcode symbols.

In fact, the experimental results demonstrated that the quality of the images reconstructed by QVGA camera resolution was comparable with those reconstructed from VGA or 1.3 megapixels, or often even better than them. As an example, Figure 7.4 illustrates images reconstructed by the three camera resolution settings, QVGA, VGA and 1.3 megapixels along with the original image.

The results of the FRR analysis are presented in Figure 7.3.

In this figure, the *peak signal to noise ratio* (PSNR) is used as a measure of the quality of reconstruction of images. The PSNR can be defined via the *mean square error* (MSE), which is a statistical measure of error used to determine the quality of images and is mathematically equivalent to the PSNR. The MSE for two $m \times n$ images I and K is defined as follows:

$$\text{MSE} = \frac{1}{mn} \sum_{i=0}^{m-1} \sum_{j=0}^{n-1} \| I\,(i,j) - K\,(i,j) \|^2.$$

The PSNR is defined as follows:

$$\text{PSNR} = 10 \, \log_{10} \left(\frac{\text{MAX}_I^2}{\sqrt{\text{MSE}}} \right) = 20 \, \log_{10} \left(\frac{\text{MAX}_I}{\text{MSE}} \right),$$

where MAX_I is the maximum pixel value of the image. When pixels are represented using eight bits, MAX_I is 255.

A higher PSNR value indicates that a reconstructed image has less noise, or greater fidelity to the original image, which in our case corresponds to a more accurate colour

		1		2		3		4		5		6		7		8		9		10	
A	QVGA	√		√		√		√		√		√		√		√		√		√	
		O	S	O	S	O	S	O	S	O	S	O	S	O	S	O	S	O	S	O	S
	VGA	√	√	√	√	√	√	√	√	√	√	√	√	√	√	√	√	√	√	√	√
	1.3 MP	√	√	√	√	√	√	√	√	√	√	√	√	√	√	√	√	√	√	√	√
B	QVGA	√		√		√		√		√		√		√		√		√		√	
		O	S	O	S	O	S	O	S	O	S	O	S	O	S	O	S	O	S	O	S
	VGA	√	√	√	√	√	√	√	√	√	√	√	√	√	√	√	√	√	√	√	√
	1.3 MP	√	√	√	√	√	√	√	√	√	√	√	√	√	√	√	√	√	√	√	√
C	QVGA	√		√		√		√		√		√		√		√		√		√	
		O	S	O	S	O	S	O	S	O	S	O	S	O	S	O	S	O	S	O	S
	VGA	√	√	√	√	√	√	√	√	√	√	√	√	√	√	√	√	√	√	√	√
	1.3 MP	√	√	√	√	√	√	√	√	√	√	√	√	√	√	√	√	√	√	√	√
D	QVGA	√		√		√		√		√		√		√		√		√		√	
		O	S	O	S	O	S	O	S	O	S	O	S	O	S	O	S	O	S	O	S
	VGA	√	√	√	√	√	√	√	√	√	√	√	√	√	√	√	√	√	√	√	√
	1.3 MP	√	√	√	√	√	√	√	√	√	√	√	√	√	√	√	√	√	√	√	√
E	QVGA	√		√		√		√		√		√		√		√		√		√	
		O	S	O	S	O	S	O	S	O	S	O	S	O	S	O	S	O	S	O	S
	VGA	√	√	√	√	√	√	√	√	√	√	√	√	√	√	√	√	√	√	√	√
	1.3 MP	√	√	√	√	√	√	√	√	√	√	√	√	√	√	√	√	√	√	√	√
F	QVGA	√		√		√		√		√		√		√		√		√		√	
		O	S	O	S	O	S	O	S	O	S	O	S	O	S	O	S	O	S	O	S
	VGA	√	√	√	√	√	√	√	√	√	√	√	√	√	√	√	√	√	√	√	√
	1.3 MP	√	√	√	√	√	√	√	√	√	√	√	√	√	√	√	√	√	√	√	√
G	QVGA	√		√		√		√		√		√		√		√		√		√	
		O	S	O	S	O	S	O	S	O	S	O	S	O	S	O	S	O	S	O	S
	VGA	√	√	√	√	√	√	√	√	√	√	√	√	√	√	√	√	√	√	√	√
	1.3 MP	√	√	√	√	√	√	√	√	√	√	√	√	√	√	√	√	√	√	√	√
H	QVGA	√		√		√		√		√		√		√		√		√		√	
		O	S	O	S	O	S	O	S	O	S	O	S	O	S	O	S	O	S	O	S
	VGA	√	√	√	√	√	√	√	√	√	√	√	√	√	√	√	√	√	√	√	√
	1.3 MP	√	√	√	√	√	√	√	√	√	√	√	√	√	√	√	√	√	√	√	√
I	QVGA	√		√		√		√		√		√		√		√		√		√	
		O	S	O	S	O	S	O	S	O	S	O	S	O	S	O	S	O	S	O	S
	VGA	√	√	√	√	√	√	√	√	√	√	√	√	√	√	√	√	√	√	√	√
	1.3 MP	√	√	√	√	√	√	√	√	√	√	√	√	√	√	√	√	√	√	√	√
J	QVGA	√		√		√		√		√		√		√		√		√		√	
		O	S	O	S	O	S	O	S	O	S	O	S	O	S	O	S	O	S	O	S
	VGA	√	√	√	√	√	√	√	√	√	√	√	√	√	√	√	√	√	√	√	√
	1.3 MP	√	√	√	√	√	√	√	√	√	√	√	√	√	√	√	√	√	√	√	√

Fig. 7.3   First-read-rate test results. One hundred MMCC™ symbols were generated, and they are labelled A1, A2, …, J10. O, original symbol; S, subsampled symbol.

reading. Visually, the images and colours captured by the higher camera resolutions appear to be sharper and clearer. However, there is no significant difference in the PSNR value. In this example the image captured by QVGA resolution achieved the best PSNR result (see Figure 7.4). In fact, in this experiment the images captured by QVGA camera often achieved better PSNR values than the higher-resolution camera images.

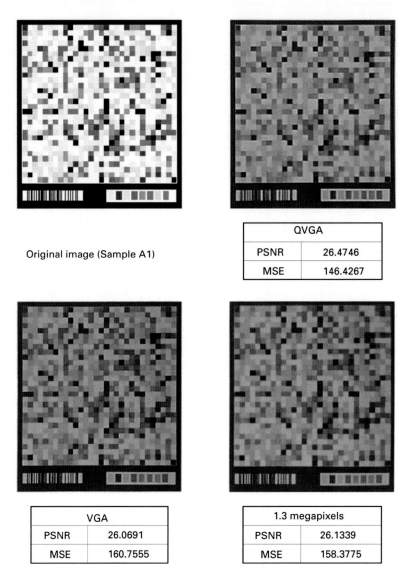

QVGA	
PSNR	26.4746
MSE	146.4267

Original image (Sample A1)

VGA	
PSNR	26.0691
MSE	160.7555

1.3 megapixels	
PSNR	26.1339
MSE	158.3775

Fig. 7.4   The resulting images after orientation and distortion correction of the captured images. The images produced by the VGA and 1.3 megapixel camera settings are reduced to the image size of QVGA. The fidelity of the reconstructed images is measured by the PSNR and MSE. For a colour version, please see Plate 18.

The three colour cluster diagrams in Figure 7.5, each of which displays 10 clusters of colours used to encode the data of the symbol in Figure 7.4, confirm the accuracy of the PSNR results. The small 'plus' symbols in grey plot the colour values of the data cells in RGB colour space. The black circle in each cluster indicates the values of the reference colour for that colour cluster (the black circles may be referred to as 'colour reference point'). The colour of each data cell is determined by calculating the distance between

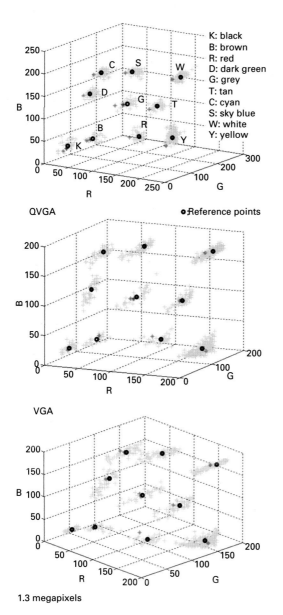

Fig. 7.5   RGB colour space plots showing the colour clusters and their separations for three different camera resolutions.

its colour values and the values of the reference colours. The colour of a data cell is considered to be the same as that of the reference cell whose colour reference point is closest.

A colour cluster of smaller extent indicates that the colour values within the cluster are closer. Consequently, each colour cluster of the original image usually appears as a thick dot. The size of the colour clusters in the QVGA image is smaller than those in the other two, and the colour separation is clearer. One possible explanation is that less noise was introduced into the colour values of the reconstructed image or else that the noise was added evenly so that it had little negative effect on accurate colour matching.

In terms of operating time, the performance of the images captured at QVGA camera resolution was significantly greater than those captured at VGA and 1.3 megapixels. The processing time for QVGA images was less than 10 seconds on average, whereas it was nearly double for the VGA images, i.e. between 15 and 20 seconds. Notably, it took between 45 and 60 seconds to process the images captured at 1.3 megapixels.

7.3.2   The results of the second experiment

In order to evaluate the robustness of the MMCC™ barcode symbols and the effectiveness of the MMCC™ decoding algorithm, a wide spectrum of potential symbol appearances was tested. These included defective symbols (i.e. a torn symbol, a smudged symbol, symbols with voids and a crumpled symbol), symbols under different illumination effects (i.e. a symbol lit by a bright light, a symbol covered by a plastic sheet and symbols under the effect of shadow), rotated symbols and distorted symbols.

All the tested symbols were successfully located and segmented, which demonstrated the robustness of the MMCC™ barcode symbol recognition algorithm. All the defective symbols except for the crumpled symbol were accurately read. This indicated the tolerance of the MMCC™ barcode to physical damage. Both the orientation and distortion correction algorithms also demonstrated their effectiveness, and the result was successful reading of all the rotated and distorted symbols.

Figures 7.6, 7.7, 7.8 and 7.9 present some examples of defective copies of the MMCC™ barcode symbol that were successfully read.

Orientation correction or distortion correction sometimes introduces a strong smear to the colour data cells of symbols. Figure 7.9(b) shows the image of sample 1 after its distortion is corrected. Owing to the smearing, the colour values of some data cells have become blended, which may confuse the MMCC™ colour-matching algorithm. The colour cluster diagram in Figure 7.9(c) displays 3D plots of the colour value sets for the data cells of sample 1 in RGB colour space. Some plotted points are observed to lie between clusters. These points may fail in the colour matching. However, Reed–Solomon codes can correct up to a certain amount of errors (22% for the samples in this experiment), resulting in a successful reading of the symbol.

In the second experiment, some limitations of the MMCC™ decoding algorithms were revealed. Figure 7.10 presents a summary of the sampled MMCC™ symbols that were not successfully read.

These include crumpled symbols captured at QVGA and VGA camera resolutions, a symbol under a strong shadow effect captured at 1.3 megapixels resolution, symbols partially under the effect of shadow and a distorted symbol with a curved edge.

Figure 7.11 presents an illegible and a legible crumpled symbol and their colour cluster diagrams. Curved lines are often introduced into crumpled symbols. Using a projective transformation, quadrilateral-to-quadrilateral mapping is performed and the distortion can be corrected. However, this does not apply to the curved lines, which often remain

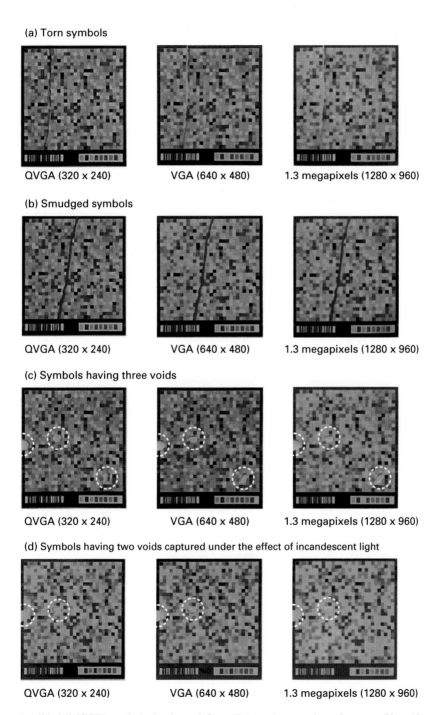

(a) Torn symbols

QVGA (320 x 240)　　　　VGA (640 x 480)　　　1.3 megapixels (1280 x 960)

(b) Smudged symbols

QVGA (320 x 240)　　　　VGA (640 x 480)　　　1.3 megapixels (1280 x 960)

(c) Symbols having three voids

QVGA (320 x 240)　　　　VGA (640 x 480)　　　1.3 megapixels (1280 x 960)

(d) Symbols having two voids captured under the effect of incandescent light

QVGA (320 x 240)　　　　VGA (640 x 480)　　　1.3 megapixels (1280 x 960)

Fig. 7.6　　　Legible MMCC™ symbols that have defects. For a colour version, please see Plate 19.

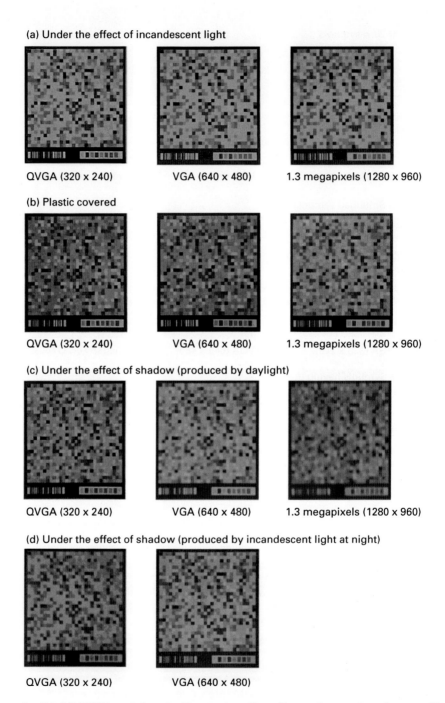

Fig. 7.7   Legible MMCC™ symbols under illumination effects. For a colour version, please see Plate 20.

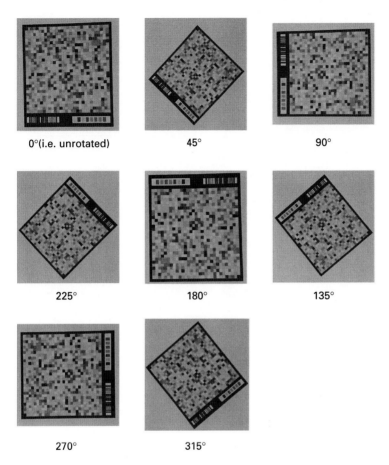

0°(i.e. unrotated)   45°   90°

225°   180°   135°

270°   315°

Fig. 7.8   Legible MMCC™ symbols in different orientations. Rotation angles are measured clockwise from the upright position. For a colour version, please see Plate 21.

as they were despite successful distortion correction based on the four corresponding points of the quadrilaterals. This causes an alignment failure, which, in turn, prevents the reading software from sampling the colour values at the correct locations, i.e. the centres of the data cells.

No significant difference is observed between the colour separations in the colour cluster diagrams of the illegible symbol and those of the legible symbols. In fact, the total number of errors for the illegible symbol exceeded the number of correctable codewords by only 2. Hence, it is possible that the reading robustness could be improved by increasing the error correction rate.

Figure 7.12 shows an illegible symbol, under a strong shadow effect, captured at 1.3 megapixels resolution. The same sample when captured at both QVGA and VGA resolutions was read successfully (see Figure 7.7). The result indicates that higher resolution of cameras is more susceptible to the lighting effect.

Figure 7.13 presents images of a sample symbol on part of which there is a shadow, captured at QVGA, VGA and 1.3 megapixels resolution. In the MMCC™ decoding

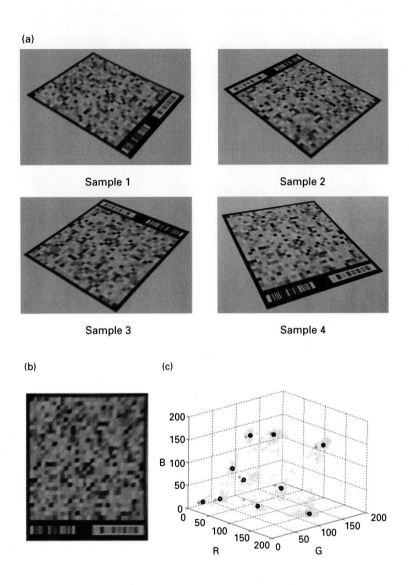

Fig. 7.9   (a) Legible MMCC™ symbols with different distortions. (b) Sample 1 after the distortion correction was performed and (c) its colour value distribution. For a colour version, please see Plate 22.

algorithm, the colour matching is performed using the colour values of each reference cell. As the examples in Figure 7.7 demonstrate, when the entire symbol including the colour reference cells is covered by a shadow the symbol is more likely to be read successfully, because not only the colour values of the data cells but also those of the reference cells are affected by the shadow. That is, the entire symbol is under the same lighting effect.

(a) Crumpled symbols

    QVGA (320 x 240)   VGA (640 x 480)

(b) Shadow cast on the MMCC™ symbol by blocking an incandescent light

  1.3 megapixels (1280 x 960)

  (c) Shadow partially cast on the MMCC™ symbol

  QVGA (320 x 240)   VGA (640 x 480)   1.3 megapixels (1280 x 960)

  (d) Distorted symbol having curved edges

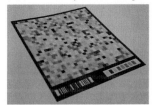

Fig. 7.10   Illegible MMCC™ symbols. For a colour version, please see Plate 23.

    In the case of the sample symbols shown in Figure 7.13, approximately half each sample symbol is under the effect of the shadow, while the other half, which includes the colour reference cells, is not.

    This causes a difference in colour values between the two halves of the symbol, although the original colours were identical. The two colour clusters within the dotted circle in Figure 7.13 are an example of such phenomena: the data cells whose colour values belong to the cluster that includes the colour reference point were under the same

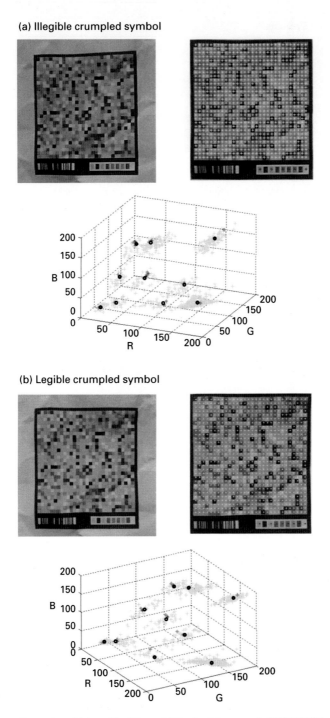

Fig. 7.11   Examples of (a) an illegible and (b) a legible crumpled MMCC™ symbol. The right-hand figures show sampled versions of the images. For a colour version, please see Plate 24.

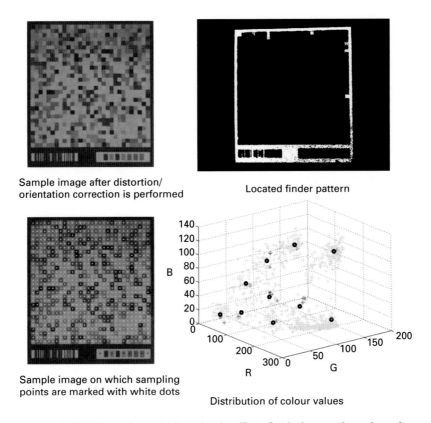

Sample image after distortion/
orientation correction is performed

Located finder pattern

Sample image on which sampling
points are marked with white dots

Distribution of colour values

Fig. 7.12   Illegible MMCC™ symbol entirely under the effect of a shadow cast by an incandescent light.
For a colour version, please see Plate 25.

lighting condition as the colour reference cells and thus were usually successfully read, whereas the other data cells often failed in the colour matching.

Figure 7.14 presents an illegible distorted symbol. When the sampled symbol was captured by the camera phone, the edge located within the white dotted line became curved. As explained previously, a projective transformation cannot correct curved lines. As a result, an alignment failure occurred. This, in turn, caused a decoding failure since the reading software failed to sample the data cell colour values at the right locations.

7.3.3   Observation

The goal of this research was to develop a colour 2D barcode that has the following advantages:

(i)   higher data density than existing 2D barcodes;
(ii) tolerance to physical damage; and
(iii) robust and reliable operation for resource-limited mobile devices.

To achieve the goal, a novel colour 2D barcode, MMCC™, was developed.

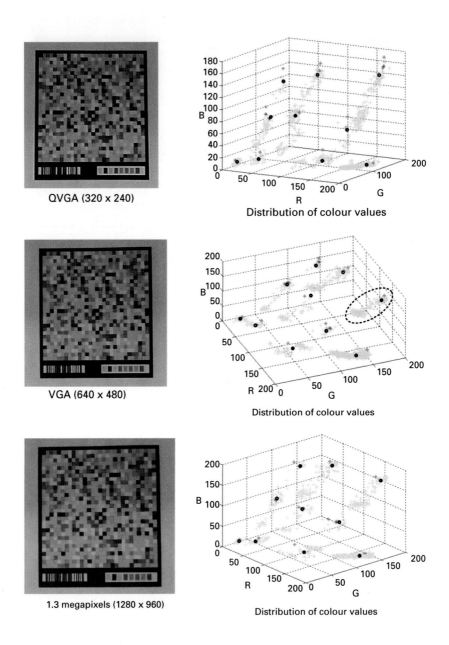

Fig. 7.13   Illegible MMCC™ symbols partially in shadow. For a colour version, please see Plate 26.

Overall performance

Throughout the experiments, the MMCC™ barcode demonstrated remarkable capability in terms of data capacity, tolerance to the different types of physical damages and reading robustness.

For the first experiment, without compression, 165 bytes of alphanumeric data with its error correction rate of approximately 22% were encoded in each 30 × 30 sample

Table 7.1. Maximum data capacity of QR Code and Data Matrix symbols roughly equivalent in symbol size to the MMCC™ sample symbols

	Symbol size	Error correction rate (%)	Data capacity (binary)
QR Code	29 × 29 cells	25	32 bytes
	33 × 33 cells	25	46 bytes
Data Matrix	26 × 26 cells	approx. 19	42 bytes
	32 × 32 cells	approx. 18	60 bytes

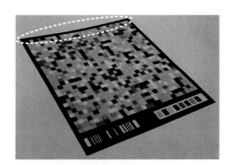

Sample image captured by
VGA camera resolution

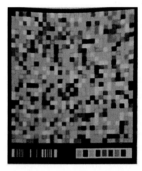

Sample image after distortion/
orientation correction

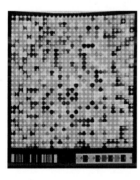

Sample image on which sampling
points are marked with white dots

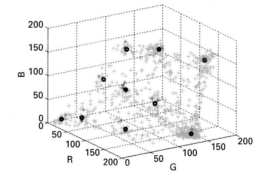

Distribution of colour values

Fig. 7.14   Illegible distorted MMCC™ symbol with a curved edge. For a colour version, please see Plate 27.

symbol. The data density of the sample symbols was nearly four times the density of industrial-standard database 2D barcodes such as QR Code and Data Matrix code (see Table 7.1).

Whereas 10 colours are used to encode the data in an MMCC™ barcode, other currently available colour database 2D barcodes such as High Capacity Color Barcode

(HCCB) and PM Code use a maximum of eight colours, or it could be even less for a camera phone implementation. Furthermore, with the compression scheme adopted for the MMCC™ barcode encoding algorithm, 11.7 kilobytes of numeric data (at an error correction rate of 22%) were encoded in the 30 × 30 sample symbol in the second experiment. Hence, as far as the data capacity is concerned, the overall achievement of the MMCC™ barcode is considered to be satisfactory.

The FRRs of all the sampled MMCC™ symbols were 100% in all three camera resolution settings. The FRRs of both the subsampled images (i.e. the images converted from 1.3 megapixels to VGA and those from VGA to QVGA) also achieved 100%. Except for some crumpled symbols, all the physically damaged symbols were successfully decoded. These results demonstrate the robustness in the MMCC™ barcode implementation and the tolerance of MMCC™ barcode to most physical damage. By increasing the error correction rate, it would be possible to improve the robustness even further.

In terms of operating time, the performance of the images produced by the lowest camera resolution was always better than that for the higher camera resolutions. The processing time for QVGA images was less than 10 seconds, on average, whereas it was nearly double for VGA images (i.e. it was between 15 and 20 seconds). The processing time for images captured at 1.3 megapixels camera resolution was rather slow, between 45 and 60 seconds. Generally, increasing the camera resolution results in an increase in the number of pixels in an image, which slows down the image processing operation.

Limitations of the MMCC™ barcode decoding algorithm

The second experiment revealed a few limitations in the MMCC™ barcode decoding algorithm. The symbols presented in Figure 7.10 were not successfully decoded. These include crumpled symbols, symbols under the effect of shadow (either partially or entirely) and distorted symbols. There are three potential factors in the decoding failure. They are:

(i) a location failure;
(ii) an alignment failure; and
(iii) a colour matching or reading failure.

A location failure often results from either a poorly designed finder pattern or unsuccessful thresholding, which is a method used to remove noise from the reconstructed image. For example, a finder pattern without distinctive features may not be located by the reading software because of background objects with similar geometric properties.

Incorrect measurement of the properties of a finder pattern (e.g. symbol size, cell size and symbol orientation) and inaccurate correction of symbol orientation or distortion cause the second type of failure, i.e. an alignment failure, even when the finder pattern has been located successfully. Acquiring the pinpoint locations of the four corners of square barcode symbols is crucial for the subsequent symbol recognition and decoding operation.

A colour matching failure may result from an alignment failure. However, it is more likely that such a failure has arisen because the colour values were sampled at the wrong location. The MMCC™ decoding algorithm is highly dependent on the values of the colour reference cells. A colour matching or reading failure may occur when the colour values of the reference cells somehow became unavailable for referencing.

We now discuss these three types of failure in more detail.

Alignment failure

Failures in reading a crumpled symbol or a distorted symbol with a curved edge result in alignment failure. In fact, MMCC™ symbol recognition and decoding algorithms were correctly performed in both cases. The problem was simply the curved edges introduced in these symbols. As explained earlier, with a projective transformation, straight lines remain straight and curved lines remain curved. The curved edge in Figure 7.14 is part of the black bounding border, i.e. the finder pattern of MMCC™ barcode, based on which symbol corrections (i.e. orientation and distortion correction) and measurement are performed. Hence, a curved edge affects all the subsequent decoding operations. The image marked with small white dots in Figure 7.14 clearly shows that some sampling points, especially at the top of the image, are out of the centres of the data cells. Incorrect measurement of the symbol size will also contribute to a failure to sample the colour reference values at the correct locations.

For both the crumpled symbols and the distorted symbols with a curved edge, the MMCC™ symbol recognition and correction procedures took place accurately as programmed. Yet the symbols were not read successfully. Hence, the only possible solution may be to develop an interactive reading software that modifies the 'capture first, decode later' approach. Generally users attempt to adjust their camera phones to read a 2D barcode symbol successfully if prompts are given interactively. This should allow the reading software to capture an image in its best state and so to avoid the reading of heavily crumpled and/or distorted images.

For crumpled symbols, no significant difference in colour separation was observed between the colour cluster diagram of the illegible crumpled symbol and that of the legible symbol. The total number of errors for the illegible symbols only slightly exceeded the number of correctable codewords. Hence, it should be possible to decode such symbols successfully by increasing the error correction rate, although this would have a negative effect in terms of actual data capacity.

Colour matching failure

The symbol under a shadow is another case of a decoding failure. The sample symbol with its top half in shadow resulted in an unsuccessful read at all three camera resolutions. Since the shadow only affected the top half of the symbol, the top half and the bottom half of the symbol are under different lighting conditions. The colour reference cells are located at the bottom of the symbol. Hence, colour matching for the data cells within the bottom half of the symbol can be accurately performed, whereas it is less likely to work effectively for the top half. The differences in colour values between the data cells under the shadow and those otherwise are presented in Figure 7.15. In this figure, differences in colour value between the data cells and the corresponding colour reference cells that are less than 30 are presented in the boxes shaded in light grey. When the difference

QVGA

Colour (Value)	Black (0 0 0)			Brown (0.5 0 0)			Red (1 0 0)			DG (0 0.5 0.5)			Grey (0.5 0.5 0.5)		
Colour component	R	G	B	R	G	B	R	G	B	R	G	B	R	G	B
Reference cell	13	16	15	48	29	29	150	39	42	28	82	92	79	93	86
Under shadow	14	14	10	41	27	16	122	31	28	32	47	34	67	62	28
	1	2	5	7	2	13	28	8	14	4	35	58	12	31	58
No shadow	18	18	16	51	24	23	152	39	30	27	79	73	80	89	73
	5	2	1	3	5	6	2	0	12	1	3	19	1	4	13
Colour (Value)	Tan (1 0.5 0.5)			Cyan (0 1 1)			SB (0.5 1 1)			White (1 1 1)			Yellow (1 1 0)		
Colour component	R	G	B	R	G	B	R	G	B	R	G	B	R	G	B
Reference cell	161	94	88	27	135	145	78	157	155	175	178	172	173	149	2
Under shadow	155	57	14	28	80	70	55	110	76	156	143	86	158	116	0
	6	37	74	1	55	75	23	47	79	19	35	86	15	33	2
No shadow	166	91	62	29	116	113	76	151	142	169	170	138	171	143	0
	5	3	26	2	19	32	2	6	13	6	8	34	2	6	2

VGA

Colour (Value)	Black (0 0 0)			Brown (0.5 0 0)			Red (1 0 0)			DG (0 0.5 0.5)			Grey (0.5 0.5 0.5)		
Colour component	R	G	B	R	G	B	R	G	B	R	G	B	R	G	B
Reference cell	20	22	24	54	38	42	142	43	57	36	81	128	79	94	126
Under shadow	16	18	14	44	24	23	129	30	28	32	50	51	68	63	53
	4	4	10	10	14	19	13	13	29	4	31	77	11	31	73
No shadow	22	23	21	59	34	36	156	38	46	33	80	118	84	86	94
	2	1	3	5	4	6	14	5	11	3	1	10	5	8	32
Colour (Value)	Tan (1 0.5 0.5)			Cyan (0 1 1)			SB (0.5 1 1)			White (1 1 1)			Yellow (1 1 0)		
Colour component	R	G	B	R	G	B	R	G	B	R	G	B	R	G	B
Reference cell	158	94	119	27	137	181	79	160	194	169	179	187	175	151	22
Under shadow	151	59	49	30	80	104	54	115	119	155	143	123	161	123	0
	7	35	70	3	53	77	25	45	75	14	36	64	14	28	22
No shadow	166	91	107	32	131	172	77	154	180	173	172	174	171	146	6
	8	3	12	5	6	9	2	6	14	4	7	13	4	5	16

1.3 megapixels

Colour (Value)	Black (0 0 0)			Brown (0.5 0 0)			Red (1 0 0)			DG (0 0.5 0.5)			Grey (0.5 0.5 0.5)		
Colour component	R	G	B	R	G	B	R	G	B	R	G	B	R	G	B
Reference cell	19	18	19	48	33	30	126	40	46	34	79	99	72	93	94
Under shadow	15	15	10	43	27	21	116	32	25	32	53	49	65	66	46
	4	3	9	5	6	9	10	8	21	2	26	50	7	24	48
No shadow	20	20	21	63	35	34	153	41	37	33	81	107	87	95	97
	1	2	2	15	2	4	27	1	9	1	2	8	15	2	3
Colour (Value)	Tan (1 0.5 0.5)			Cyan (0 1 1)			SB (0.5 1 1)			White (1 1 1)			Yellow (1 1 0)		
Colour component	R	G	B	R	G	B	R	G	B	R	G	B	R	G	B
Reference cell	154	84	104	15	131	165	71	156	181	161	172	178	165	148	1
Under shadow	151	62	39	26	87	94	54	120	109	155	146	119	161	122	0
	3	22	65	11	44	71	17	36	72	6	26	59	4	26	1
No shadow	166	94	88	29	131	162	78	150	168	174	174	176	174	149	0
	12	10	16	14	0	3	7	6	13	13	2	2	9	1	1

The difference from the value of the corresponding reference colour is less than 30.

The difference from the value of the corresponding reference colour is greater than 30.

Fig. 7.15   Effect of a partial shadow on the colour values. Five samples of each colour under each condition (either under the effect of shadow or otherwise) were randomly selected from the images captured at QVGA, VGA and 1.3 megapixels camera resolution. Then, the mean values of the R, G, B colour components for each colour under each condition (e.g. the image under the shadow captured by QVGA camera resolution) were calculated.

is greater than or equal to 30 in the assumed range [0, 255], the boxes are shaded in a darker grey. The differences between the data cells without the shadow effect and the corresponding colour reference cells are quite small for all the 10 colours, resulting in successful colour matching and reading. Conversely, the differences between the data cells under the shadow and the corresponding colour reference cells are rather large, which led to the colour matching failure.

The MMCC™ barcode decoding algorithm is based on the use of colour reference cells. Hence, successful colour reading is highly dependent on reliable reference colour values. When the data area is partly or entirely under a different lighting effect from the colour reference area, it is less likely that the values in the reference cells will provide the information necessary for colour matching of the data cells.

Possible solutions for this type of problem are either allocating the reference area to more than one location or developing an effective colour thresholding method so that the lighting effect can be removed. However, the former solution may not cover all possible scenarios. Furthermore, it would reduce the data capacity of the MMCC™ barcode for a given space. While we were developing the MMCC™ barcode decoding algorithm, the effectiveness of several local thresholding methods was examined. The outcomes were not very promising since it was often the case that colours with similar values were confused with each other. However, it may be worthwhile investigating the latter possibility further.

An observation worthy of special mention is that the green and blue components in RGB colour space are more susceptible to a shadow or lighting effect than the red component. Figure 7.15 shows, in the 'under shadow' rows, that the colour values of the G and B components decreased significantly for all the images captured at QVGA, VGA and 1.3 megapixels. The susceptibility of the B component is markedly high and its values under the shadow effect were often less than half those without shadow. However, only insignificant differences in the values of the R component were observed between the cells with and without the shadow effect. In fact, few significant differences were observed, between the data cells under the shadow and the cells without shadow, in the values of black, brown and red colours in which no G and/or B components were present.

This confirms the effectiveness of the proposed colour selection scheme, in which nine colours are selected in the plane in the RGB colour cube that has black, red, white and cyan as its four corners. Since the plane does not include either blue or green, the colours on the plane are less susceptible to the lighting effect. This, in turn, contributes to the reading robustness of MMCC™ barcodes.

Camera resolution and image fidelity

In addition to the above two cases of failure, the sample image under the dark shadow captured at 1.3 megapixels was also illegible, although the images of the identical sample captured at both the QVGA and VGA resolutions were legible. The image marked with white dots and the colour cluster diagram in Figure 7.12 indicate that the decoding failure results from a colour matching failure rather than an alignment failure. Although the colour sampling was performed at the correct locations, numerous sets of colour values were out of the range of every colour cluster. This indicates that the colour values of the symbol have changed, having been strongly affected by the shadow. The result is incorrect colour matching.

It is widely believed that higher-resolution cameras enable pinpoint accuracy of an object's location and, as a result, produce better-quality images. This is an advantage in representing a scene accurately as it is, including both the main objects and the effects on the objects (e.g. illumination, shadow etc.). However, it is possible that the accuracy in the reconstructed image produced by a higher camera resolution has not only positive features (e.g. sharpness) but also negative features (e.g. shadow) in barcode reading. This may explain why images captured at 1.3 megapixels resolution were more sensitive to lighting effects such as shadow and under these conditions were sometimes not read successfully.

Another possible explanation is that the quality of the images captured at 1.3 mega-pixels resolution was not as good as those of QVGA and VGA, despite the general belief otherwise. According to [63], a higher camera resolution does not necessarily mean better-quality images are produced. This is especially true when a charge coupled device (CCD) image sensor is implemented using the interline transfer CCD (IT-CCD) architecture presented in Figure 7.16, which is standard for current digital mobile phone cameras.

A pixel is the basic unit of a digitised image. If the camera resolution is 1.3 megapixels, the surface of the CCD is divided into 1 300 000 pixels. The source of digital data is light. To create digital image data, first the CCD converts light captured through the photodiode into electric charges, which are then gathered and sent to an amplifier through the vertical CCD (VCCD) and the horizontal CCD (HCCD). That is, each pixel is divided into two parts in the IT-CCD architecture, one part converting light into electric charges (i.e. the photodiode) and the other transferring them to the amplifier (i.e. the VCCD).

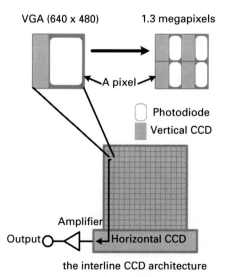

Fig. 7.16   The effect on the interline CCD architecture when the resolution increases from VGA to 1.3 megapixels (adapted from [74]).

When the camera resolution is increased, for example from VGA (640 × 480) to 1.3 megapixels (1280 × 960), without changing the size of the CCD, the sizes of both the photodiode and the VCCD of each pixel will become one quarter of their original values. The problem is that we cannot reduce the size of the VCCD as far as one quarter of its original size and hence the size of the photodiode suffers (see Figure 7.16): it becomes even smaller than one quarter of its original size, which in turn reduces the surface available to capture light and thus degrades the quality of the captured image. Hence, increasing the camera resolution may enable pinpoint accuracy but, at the same time, it has a negative effect on the efficient capture of light in the IT-CCD architecture. This may debunk the myth that the performance of higher-resolution cameras is always better than those of lower resolution.

7.3.4 Key findings

Appropriate camera resolution for robust barcode reading

It is widely believed that cameras with higher resolution can produce better quality in images and, consequently, increase the robustness in barcode reading. However, in fact it is not desirable to use higher camera resolutions for 2D barcode reading, for the following two reasons.

(i) Higher-resolution camera images are more susceptible to the lighting effect than lower-resolution images.
(ii) The processing time for images captured at a higher camera resolution is significantly greater than that for images captured at a lower camera resolution.

In our experiments, the camera resolution of the mobile phone (i.e. a Sanyo W33SA II) was set to QVGA (i.e. 320 × 240 pixels), VGA (i.e. 640 × 480 pixels) or 1.3 megapixels (i.e. 1280 × 960 pixels). The experimental results indicate that a higher camera resolution does not guarantee better barcode reading. In fact, the performances at lower camera resolutions (i.e. QVGA and VGA) were better than that of the highest camera resolution (i.e. 1.3 megapixels) in terms of both operating time and reading robustness (e.g. clearness in colour separation and tolerance of the lighting effect).

Since we did not include camera resolutions that are either less than QVGA or greater than 1.3 megapixels in the experiments, it might be too early to reach a definite conclusion. However, considering the availability of camera phones and their processing time and reading robustness, either QVGA or VGA camera resolutions might be a good choice for robust and fast barcode reading. In practice, the auto-focus function is more important than high camera resolution. It enables precise measurement of the size of a barcode symbol and this, in turn, increases accuracy in correcting orientation and/or distortion when locating sampling points and reading the symbol.

Flexibility of an interactive reading software

As a result of our experiments, two limitations in the MMCC™ symbol recognition and code decoding algorithms have been revealed: a difficulty in decoding symbols with curved edges and an inaccuracy in the reading of colours when the data cells and the colour reference cells are under different lighting conditions (e.g. when a symbol is partially covered by shadow). It is possible to overcome these problems by developing an interactive reading software. In general, users attempt to adjust their camera phones to read 2D barcode symbols successfully if prompts are given interactively. This provides the reading software with flexibility and allows it to capture an image in its best state and avoids the reading of heavily crumpled and/or distorted images.

Robust and fast operation with a subsampling approach

The image processing time depends on the number of pixels in an image. The images captured at higher camera resolutions have a larger size, which slows down the processing time. The subsampling approach enables faster reading operation. In the experiments, the performance of the subsampled images was comparable with that of the original images in terms of their FRRs. Hence the subsampling approach may help to improve the overall performance of MMCC™ barcode reading. For example, when a reconstructed image is too small to be accurately read, capturing the image at a higher camera resolution (e.g. VGA) and reading its subsampled image (at e.g. QVGA) might be effective in terms of both reading robustness and operating time. This approach, with the use of a zooming function, may also help to improve the reading distance of MMCC™ barcodes.

7.4   Conclusions and future work

In order to evaluate the effectiveness and robustness of the MMCC™ symbol-recognition algorithm and decoding algorithm, we conducted two types of experiment, namely, FRR analysis and an analysis of reading robustness.

Overall, the MMCC™ barcode demonstrated remarkable capability in terms of data capacity, tolerance to various types of physical damage and reading robustness. The reading robustness, in turn, verifies the effectiveness and stability of the MMCC™ symbol-recognition and decoding algorithms. It also confirms the effectiveness of the proposed colour selection scheme, which aims to maximise the separation of the colours used for encoding data across more than one colour space.

Although during the experiments the two limitations mentioned above were revealed in the MMCC™ symbol-recognition and code decoding algorithms, it is possible to overcome these problems by developing an interactive reading software. Consequently, a worthy challenge for future work is the development of an interactive reading software that not only overcomes the issues stated above but also enables fast and accurate decoding of MMCC™ barcodes.

7.5   Summary

The MMCC™ symbology was developed with the aim of inventing a novel colour 2D barcode that satisfies the following requirements:

(i)   higher data capacity than any existing 2D barcodes;
(ii) tolerance to physical damage; and
(iii) robust and reliable operation for resource-limited mobile devices.

In order to verify the effectiveness and robustness of the MMCC™ barcode and its recognition and decoding algorithms, two types of experiments were conducted:

 (i) an examination of the overall performance of MMCC™ barcode; and
(ii) an analysis of the reading robustness.

To evaluate the overall performance, the FRRs of sample symbols were analysed, while the reading robustness was measured by decoding copies of a sample symbol in various conditions of defectiveness.

Overall, the MMCC™ barcode has demonstrated remarkable capability in terms of data capacity, tolerance to different types of physical damage and reading robustness. The FRRs of all the sampled MMCC™ symbols were 100% in all three camera resolution settings. The FRRs of both subsampled images (i.e. the images converted from 1.3 megapixels to VGA and those from VGA to QVGA) also achieved 100%. Except for some crumpled symbols, all the physically damaged symbols were successfully decoded. These results demonstrate the robustness in MMCC™ barcode implementation and the tolerance of MMCC™ barcode towards likely types of physical damage.

Through these experiments, we have also identified ways to further improve the effectiveness and robustness of the MMCC™ symbol-recognition and code decoding algorithms. The three key findings are as follows:

  (i) An appropriate camera resolution is desirable for robust barcode reading.
 (ii) The flexibility of interactive reading software is needed.
(iii) Fast and robust operation is possible with the subsampling approach.

Although the experimental results reveal two limitations in the MMCC™ symbol-recognition and decoding algorithms, it should be possible to overcome these problems by developing an interactive reading software instead of the current 'capture first, decode later' approach. This will be further investigated as a future work.

Appendix A   Barcode applications

Table A.1. Current use of barcode technology – (i)

Category	Purpose or object of barcode use
Commerce	anti-theft, publications, books (bookshop, price code), magazines (UPC, supplementary), library (inventory, shelf, checkout), video stores, music, catalogues, turnpikes, highway control, banking (ATM, money recodes, cheques and documents, money and inventory)
Industry	quality control (verification of correct parts, trackability of components, finished assembly), materials (management, material flow, stock control, inventory, tracking material), production control, control machinery, robotics, install components, photographic, plant operations, electronic (components and production), paper production, material resource planning and similar systems, advance shipping notice, electronic proof of shipping, electronic proof of receipt, automotive bodies (paint, parts control and others), computer boards and many production operations, light industry and job shops (engineering, design, documents, pre-production models and parts tools)
Retail	checkout, ordering, receipt, damaged, bad stock, inventory, control, self-checkout (research and development), purchasing

Table A.2. Current use of barcode technology – (ii)

Category	Purpose or object of barcode use
Distribution	packaging (primary, secondary, shipping container, unit loads, others), trucks and rail (cars or wagons, barcodes or transducer sensors), package delivery (continuous control, customer satisfaction, hub operation, detailed information to ensure overnight delivery), postal codes, conveyors, warehousing, truck loading, vehicle location, material movement, total data information
Medical	patient history (transportable medical record), smart card (matrix, stacked code), equipment, patients, staff authorisation, pharmaceuticals, unit dose, other patient-used items to be charged, human blood, other materials with life or death effects, controlling shelf life, organisation of standards, HIBCC (Health Industry Business Communications Council), NWDA (National Wholesale Drug Association), European medical organisations
Documents	forms, sales and marketing, order forms, shipping and receiving, accounting, payroll, scheduling, vacation
Other	traffic and sensors, person tracking and location, vehicle tracking and location, transponders, security, card (eye print, voice, other), voice recognition, research and development, tracking items and tests, traceability, radio and TV cassettes, police inspections, emission inspections, kidnapped or lost children, Mount Everest equipment control, sports (long distance runners, Olympics: access and security), bees, fish and other animals, ubiquitous computing (Web link, e-ticket or e-coupon, plasma display control)

Appendix B Automatic identification technologies

	Type	Image	Description and use
1	Barcode	Bar Code	Dark rectangular bars and white rectangular spaces are alternated, the widths of individual rectangles are varied to code information.
2	Magnetic stripe	 MRC	Magnetic cards, prepaid cards, hybrid IC-magnetic cards, etc.
3	OCR	 OCR	Optical character recognition.
4	RF/DC (radio frequency data identification)	 RFID	Identification device for radio frequency recording of data on card or tag-shaped media and for communications through an antenna.
5	RF/DC (radio frequency data communications)	 RFDC	Generic term for communications systems involving two-way communications between a host computer and portable terminals for collecting and transmitting or receiving data.

Fig. B.1   Types of automatic identification (AI) technologies – (i) (adapted from [13]).

6	Smart card		Credit card with built-in CPU and memory. Has superior security function and communications function for data recording and retrieval.
7	Optical card		Credit-card size ROM with optically recorded and stored large amount of data.
8	Memory button (tough memory)		Essentially functions the same way as RF/ID. Tag or card-shaped and has a stainless steel cover. Data are transferred through contact with this cover.
9	Biometric identification	 Biometrics	Identification by biometric characteristics such as finger prints or retinal patterns.
10	Machine vision	 Vision	Image processing by computer.
11	Voice recognition	 Auto ID Voice	Converts human voice into electric signal and inputs them into computer.
12	EDI (electronic data interchange)		Online transmission and receipt of all transaction data including customer name, product name and price information to facilitate processing steps.

Fig. B.2   Types of AI technologies – (ii) (adapted from [13]).

Appendix C Barcode history timeline

Table C.1. Barcode history timeline – (i)

Year	Event
1932	A business student named Wallace Fling wrote a Master's thesis on automatic identification for supermarkets using punched cards.
1949	Norman Joseph Woodland and Bernard Silver filed a patent application entitled 'Classifying apparatus and method'.
1952	The patent application submitted by Woodland and Silver was issued as US Patent 2 612 994.
1962	Prototype barcode and scanning systems to identify railroad cars were developed.
1967	RCA was installed as one of the first scanning systems at a Kroger store in Cincinnati. The railroads in North America adopted the tracking system developed by Sylvania/GTE.
1969	NAFC asked Logicon Inc. to develop a proposal for an industry-wide barcode system. Computer Identics developed HeNe laser scanner. The CCD was invented by Willard Boyle and George E. Smith at AT&T Bell Laboratories.
1970	Universal Grocery Products Identification Code (UGPIC) was announced.

Table C.2. Barcode history timeline – (ii)

Year	Event
1971	The 10-digit numeric code was selected.
	IBM developed Delta Distance Code.
	Plessey Code was developed by Plessey Co.
1972	The Uniform Code Council (UCC) was established.
1973	The Uniform Grocery Products Code Council (UGPCC) announced the UPC symbol and code.
1974	One of the first UPC scanners, made by National Cash Register Co. was installed at Marsh supermarket in Troy and Ohio for the regular use of UPC symbols.
	Code 39, the first practical alphanumeric barcode, was invented by David Allais and Ray Stevens of Intermec Co.
1977	The European Article Numbering Association was established and the EAN barcode format was adopted.
	The US DoD commenced a research project on barcodes called LOGMARS.
1981	Computer Identics developed Code 128.
	The US DoD adopted the use of Code 39 for marking all products sold to the US military.
1982	Symbol Technologies Inc. developed the hand HeNe laser scanner.
1984	The Automotive Industry Action Group (AIAG) adopted Code 39 as a standard label.
1987	The Energy Information Administration (EIA) adopted Code 39 as a standard label.

Appendix D   MicroPDF417 data capacity and error correction capability

Table D.1. MicroPDF417 data capacity and error correction capability for column versions 1 and 2 (adapted from [13])

Column number (version)	Row number	Correctable codeword	Numeric compaction mode	Text compaction mode	Binary compaction mode
1	11	7	8	6	3
	14	7	17	12	7
	17	7	26	18	10
	20	8	32	22	13
	24	8	44	30	18
	28	8	55	38	22
2	8	8	20	14	8
	11	9	35	24	14
	14	9	52	36	21
	17	10	67	46	27
	20	11	82	56	33
	23	13	93	64	38
	26	15	105	72	43

Table D.2. MicroPDF417 data capacity and error correction capability for column versions 3 and 4 (adapted from [13])

Column number (version)	Row number	Correctable codeword	Numeric compaction mode	Text compaction mode	Binary compaction mode
3	6	12	14	10	6
	8	14	26	18	10
	10	16	38	26	15
	12	18	49	34	20
	15	21	67	46	27
	20	26	96	66	39
	26	32	132	90	54
	32	38	167	114	68
	38	44	202	138	82
	44	50	237	162	97
4	4	8	20	14	8
	6	12	32	22	13
	8	14	49	34	20
	10	16	67	46	27
	12	18	85	58	34
	15	21	111	76	45
	20	26	155	106	63
	26	32	208	142	85
	32	38	261	178	106
	38	44	313	214	128
	44	50	366	250	150

Appendix E QR Code symbol version and data capacity

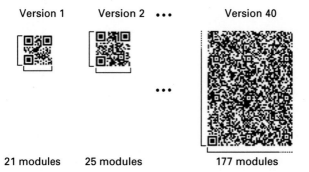

Version 1   Version 2 •••   Version 40

•••

21 modules   25 modules   177 modules

Fig. E.1   QR Code symbol versions (adapted from [24]), showing the numbers of modules (cells) in a symbol.

Table E.1. Maximum QR Code data capacity for versions 1 to 10

Version	Modules	ECC level	Data bits	Numeric	Alphanumeric	Binary	Kanji
1	21 × 21	L	152	41	25	17	10
		M	128	34	20	14	8
		Q	104	27	16	11	7
		H	72	17	10	7	4
2	25 × 25	L	272	77	47	32	20
		M	224	63	38	26	16
		Q	176	48	29	20	12
		H	128	34	20	14	8
3	29 × 29	L	440	127	77	53	32
		M	352	101	61	42	26
		Q	272	77	47	32	20
		H	208	58	35	24	15
4	33 × 33	L	640	187	114	78	48
		M	512	149	90	62	38
		Q	384	111	67	46	28
		H	288	82	50	34	21
5	37 × 37	L	864	255	154	106	65
		M	688	202	122	84	52
		Q	496	144	87	60	37
		H	368	106	64	44	27
6	41 × 41	L	1088	322	195	134	82
		M	864	255	154	106	65
		Q	608	178	108	74	45
		H	480	139	84	58	36
7	45 × 45	L	1248	370	224	154	95
		M	992	293	178	122	75
		Q	704	207	125	86	53
		H	528	154	93	64	39
8	49 × 49	L	1552	461	279	192	118
		M	1232	365	221	152	93
		Q	880	259	157	108	66
		H	688	202	122	84	52
9	53 × 53	L	1856	552	335	230	141
		M	1456	432	262	180	111
		Q	1056	312	189	130	80
		H	800	235	143	98	60
10	57 × 57	L	2192	652	395	271	167
		M	1728	513	311	213	131
		Q	1232	364	221	151	93
		H	976	288	174	119	74

Table E.2. Maximum QR Code data capacity for versions 11 to 20

Version	Modules	ECC level	Data bits	Numeric	Alphanumeric	Binary	Kanji
11	61×61	L	2592	772	468	321	198
		M	2032	604	366	251	155
		Q	1440	427	259	177	109
		H	1120	331	200	137	85
12	65×65	L	2960	883	535	367	226
		M	2320	691	419	287	177
		Q	1648	489	296	203	125
		H	1264	374	227	155	96
13	69×69	L	3424	1022	619	425	262
		M	2672	796	483	331	204
		Q	1952	580	352	241	149
		H	1440	427	259	177	109
14	73×73	L	3688	1101	667	458	282
		M	2920	871	528	362	223
		Q	2088	621	376	258	159
		H	1576	468	283	194	120
15	77×77	L	4184	1250	758	520	320
		M	3320	991	600	412	254
		Q	2360	703	426	292	180
		H	1784	530	321	220	136
16	81×81	L	4712	1408	854	586	361
		M	3624	1082	656	450	277
		Q	2600	775	470	322	198
		H	2024	602	365	250	154
17	85×85	L	5176	1548	938	644	397
		M	4056	1212	734	504	310
		Q	2936	876	531	364	224
		H	2264	674	408	280	173
18	89×89	L	5768	1725	1046	718	442
		M	4504	1346	816	560	345
		Q	3176	948	574	394	243
		H	2504	746	452	310	191
19	93×93	L	6360	1903	1153	792	488
		M	5016	1500	909	624	384
		Q	3560	1063	644	442	272
		H	2728	813	493	338	208
20	97×97	L	6888	2061	1249	856	528
		M	5352	1600	970	666	410
		Q	3880	1159	702	482	297
		H	3080	919	557	382	235

Table E.3. Maximum QR Code data capacity for versions 21 to 30

Version	Modules	ECC level	Data bits	Numeric	Alphanumeric	Binary	Kanji
21	101 × 101	L	7456	2232	1352	929	572
		M	5712	1708	1035	711	438
		Q	4096	1224	742	509	314
		H	3248	969	587	403	248
22	105 × 105	L	8048	2409	1460	1003	618
		M	6256	1872	1134	779	480
		Q	4544	1358	823	565	348
		H	3536	1056	640	439	270
23	109 × 109	L	8752	2620	1588	1091	672
		M	6880	2059	1248	857	528
		Q	4912	1468	890	611	375
		H	3712	1108	672	461	284
24	113 × 113	L	9392	2812	1704	1171	721
		M	7312	2188	1326	911	561
		Q	5312	1588	963	661	407
		H	4112	1228	744	511	315
25	117 × 117	L	10208	3057	1853	1273	784
		M	8000	2395	1451	997	614
		Q	5744	1718	1041	715	440
		H	4304	1286	779	535	330
26	121 × 121	L	10960	3283	1990	1367	842
		M	8496	2544	1542	1059	652
		Q	6032	1804	1094	751	462
		H	4768	1425	864	593	365
27	125 × 125	L	11744	3514	2132	1465	902
		M	9024	2701	1637	1125	692
		Q	6464	1933	1172	805	496
		H	5024	1501	910	625	385
28	129 × 129	L	12248	3669	2223	1528	940
		M	9544	2857	1732	1190	732
		Q	6968	2085	1263	868	534
		H	5288	1581	958	658	405
29	133 × 133	L	13048	3909	2369	1628	1002
		M	10136	3035	1839	1264	778
		Q	7288	2181	1322	908	559
		H	5608	1677	1016	698	430
30	137 × 137	L	13880	4158	2520	1732	1066
		M	10984	3289	1994	1370	843
		Q	7880	2358	1429	982	604
		H	5960	1782	1080	742	457

Table E.4. Maximum QR Code data capacity for versions 31 to 40

Version	Modules	ECC level	Data bits	Numeric	Alphanumeric	Binary	Kanji
31	141 × 141	L	14744	4417	2677	1840	1132
		M	11640	3486	2113	1452	894
		Q	8264	2473	1499	1030	634
		H	6344	1897	1150	790	486
32	145 × 145	L	15640	4686	2840	1952	1201
		M	12328	3693	2238	1538	947
		Q	8920	2670	1618	1112	684
		H	6760	2022	1226	842	518
33	149 × 149	L	16568	4965	3009	2068	1273
		M	13048	3909	2369	1628	1002
		Q	9368	2805	1700	1168	719
		H	7208	2157	1307	898	553
34	153 × 153	L	17528	5253	3183	2188	1347
		M	13800	4134	2506	1722	1060
		Q	9848	2949	1787	1228	756
		H	7688	2301	1394	958	590
35	157 × 157	L	18448	5529	3351	2303	1417
		M	14496	4343	2632	1809	1113
		Q	10288	3081	1867	1283	790
		H	7888	2361	1431	983	605
36	161 × 161	L	19472	5836	3537	2431	1496
		M	15312	4588	2780	1911	1176
		Q	10832	3244	1966	1351	832
		H	8432	2524	1530	1051	647
37	165 × 165	L	20528	6153	3729	2563	1577
		M	15936	4775	2894	1989	1224
		Q	11408	3417	2071	1423	876
		H	8768	2625	1591	1093	673
38	169 × 169	L	21616	6479	3927	2699	1661
		M	16816	5039	3054	2099	1292
		Q	12016	3599	2181	1499	923
		H	9136	2735	1658	1139	701
39	173 × 173	L	22496	6743	4087	2809	1729
		M	17728	5313	3220	2213	1362
		Q	12656	3791	2298	1579	972
		H	9776	2927	1774	1219	750
40	177 × 177	L	23648	7089	4296	2953	1817
		M	18672	5596	3391	2331	1435
		Q	13328	3993	2420	1663	1024
		H	10208	3057	1852	1273	784

Appendix F   Data Matrix data capacity and error correction capability

Table F.1. Data Matrix data capacity and error correction – (i) Rectangular symbol (adapted from [13])

Symbol size		Data region		Codeword no.		Maximum data capacity			EC/data+EC[d]
Row	Column	Size	Block no.	Data	EC	N^a	A^b	B^c	(%)
8	18	6×16	1	5	7	10	6	3	58.3
8	32	6×14	2	10	11	20	13	8	52.4
12	26	10×20	1	16	14	32	22	14	46.7
12	36	10×16	2	22	18	44	31	20	45.0
16	36	14×16	2	32	24	64	46	30	42.9

[a] Numeric.
[b] Alphanumeric.
[c] Eight-bit byte.
[d] The number of error correction bits divided by the sum of this number and the number of data bits.

Table F.2. Data Matrix data capacity and error correction – (ii) Square symbol (adapted from [13])

Symbol size		Data region		Codeword no.		Maximum data capacity			ECC[d]	EC/data+EC
Row	Column	Size	Block no.	Data	EC	N[a]	A[b]	B[c]	(%)	(%)
10	10	8 × 8	1	3	5	6	3	1	25	62.5
12	12	10 × 10	1	5	7	10	6	3	25	58.3
14	14	12 × 12	1	8	10	16	10	6	28–39	55.6
16	16	14 × 14	1	12	12	24	16	10	25–38	50.0
18	18	16 × 16	1	18	14	36	25	16	22–34	43.8
20	20	18 × 18	1	22	18	44	31	20	23–38	45.0
22	22	20 × 20	1	30	20	60	43	28	20–34	40.0
24	24	22 × 22	1	36	24	72	52	34	20–35	40.0
26	26	24 × 24	1	44	28	88	64	42	19–35	38.9
32	32	14 × 14	4	62	36	124	91	60	18–34	36.7
36	36	16 × 16	4	86	42	172	127	84	16–30	32.8
40	40	18 × 18	4	114	48	228	169	112	15–28	29.6
44	44	20 × 20	4	144	56	288	214	142	14–27	28.0
48	48	22 × 22	4	174	68	348	259	172	14–27	28.1
52	52	24 × 24	4	204	84	408	304	202	15–27	29.2
64	64	14 × 14	16	280	112	560	418	278	14–27	28.6
72	72	16 × 16	16	368	144	736	550	366	14–26	28.1
80	80	18 × 18	16	456	192	912	682	454	15–28	29.6
88	88	20 × 20	16	576	224	1152	862	574	14–27	28.0
96	96	22 × 22	16	696	272	1392	1042	694	14–27	28.1
104	104	24 × 24	16	816	336	1632	1222	814	15–28	29.2
120	120	18 × 18	36	1050	408	2100	1573	1048	14–27	28.0
132	132	20 × 20	36	1304	496	2608	1954	1302	14–26	27.6
144	144	22 × 22	36	1558	620	3116	2335	1556	14–27	28.5

[a] Numeric.
[b] Alphanumeric.
[c] Eight-bit byte.
[d] The data restoration rates for the misread symbols and those that could not be read at the first attempt are different; hence, the whole range (e.g. 28–39) is presented.

Appendix G   An image under the every-other-line effect

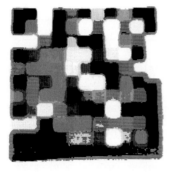

Fig. G.1   An image under the every-other-line effect due to the zigzag scanning of an adaptive thresholding method. For a colour version, please see Plate 28.

Appendix H Colour susceptibility and distance between colours

Distance between colours in the RGB cube

Original colour: range (0 – 255)								
Colour	Black	Blue	Green	Cyan	Red	Magenta	Yellow	White
Black (K)	0	57.4565	86.3061	104.6016	115.7476	121.0382	171.2119	198.9441
Blue (B)	57.4565	0	57.8835	50.7272	121.2714	114.7933	151.9498	161.6215
Green (G)	86.3061	57.8835	0	50.1523	105.7934	102.8081	102.8117	124.5873
Cyan (C)	104.6016	50.7272	50.1523	0	137.1915	125.5080	132.7987	128.2819
Red (R)	115.7476	121.2714	105.7934	137.1915	0	27.1155	100.6578	137.5473
Magenta (M)	121.0382	114.7933	102.8081	125.5080	27.1155	0	97.7305	120.9111
Yellow (Y)	171.2119	151.9498	102.8117	132.7987	100.6578	97.7305	0	68.6822
White (W)	198.9441	161.6215	124.5873	128.2819	137.5473	120.9111	68.6822	0

 Distance less than 60

Distance between colours within each plane

		4 corner points	Average distance	Minimum distance
Red axis	KRWC	137.0000	104.6016
	BMYG	105.0000	57.8835
Green axis	RYCB	116.0000	50.7272
	KGWM	126.0000	86.3061
Blue axis	KBWY	135.5000	57.4565
	RMCG	91.4000	27.1154

Probability of successful read

Black	100.00
Blue	77.78
Green	96.30
Cyan	100.00
Red	100.00
Magenta	70.37
Yellow	85.19
White	88.89

Fig. H.1 Experimental results (i). Distances between colours in the RGB cube and within each plane. The lower right panel shows the probability of a successful read; the colour values have each been normalised by the average of all the low and high values.

Printed symbols captured by Nokia 6110

Original colour: range [0, 1]				Red [0, 255]			Green [0, 255]			Blue [0, 255]		
Colour	R	G	B	Min	Max	Mean	Min	Max	Mean	Min	Max	Mean
Black (K)	0	0	0	0	18	3	0	18	6	0	9	1
Blue (B)	0	0	1	0	11	0	14	41	30	37	73	57
Green (G)	0	1	0	6	46	25	71	89	80	3	59	33
Cyan (C)	0	1	1	0	16	1	60	87	74	68	111	90
Red (R)	1	0	0	104	139	121	21	55	39	0	49	20
Magenta (M)	1	0	1	106	149	124	14	60	38	24	30	47
Yellow (Y)	1	1	0	113	137	124	116	138	126	10	63	38
White (W)	1	1	1	108	131	118	114	133	124	97	127	111

Printed symbols captured by Nokia 6300

Original colour: range [0, 1]				Red [0, 255]			Green [0, 255]			Blue [0, 255]		
Colour	R	G	B	Min	Max	Mean	Min	Max	Mean	Min	Max	Mean
Black (K)	0	0	0	0	23	10	4	25	16	0	8	1
Blue (B)	0	0	1	0	16	5	27	48	38	32	63	50
Green (G)	0	1	0	21	47	35	77	97	87	34	63	50
Cyan (C)	0	1	1	0	23	11	66	83	74	67	88	79
Red (R)	1	0	0	102	133	117	6	42	22	1	44	20
Magenta (M)	1	0	1	94	128	110	1	39	20	27	68	47
Yellow (Y)	1	1	0	108	132	122	117	139	127	46	78	62
White (W)	1	1	1	110	126	118	120	138	129	116	137	126

Symbols displayed on the LCD screen captured by Nokia 6110

Original colour: range [0, 1]				Red [0, 255]			Green [0, 255]			Blue [0, 255]		
Colour	R	G	B	Min	Max	Mean	Min	Max	Mean	Min	Max	Mean
Black (K)	0	0	0	0	27	3	0	38	18	0	50	17
Blue (B)	0	0	1	27	80	55	109	135	121	167	199	183
Green (G)	0	1	0	67	96	82	157	175	167	57	92	75
Cyan (C)	0	1	1	57	95	79	156	182	169	145	174	160
Red (R)	1	0	0	161	188	176	69	96	82	35	76	58
Magenta (M)	1	0	1	151	177	163	109	127	118	149	180	165
Yellow (Y)	1	1	0	149	175	164	148	168	158	84	117	98
White (W)	1	1	1	143	168	154	142	160	152	131	164	150

Fig. H.2   Experimental results (ii). The susceptibility of the RGB primary and secondary colours.

Appendix I   MMCC™ encoding and decoding

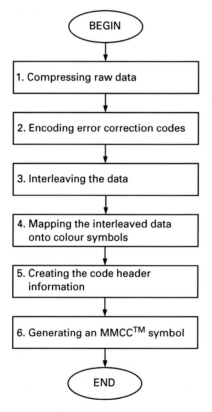

Fig. I.1   A flowchart illustrating the method of encoding an MMCC™ barcode, according to the encoding algorithm.

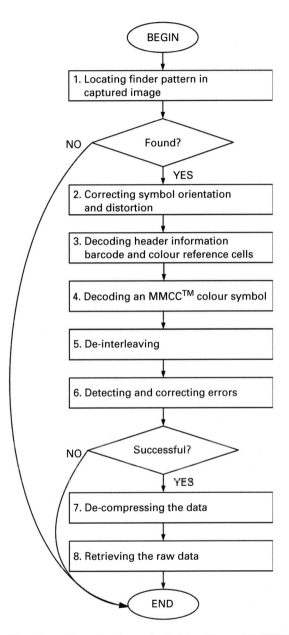

Fig. I.2   Flowcharts illustrating the method of decoding an MMCC™ barcode, according to the decoding algorithm.

Appendix J   First-read-rate test samples

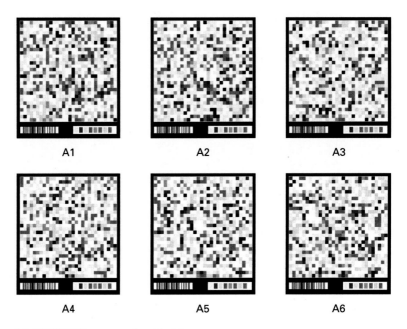

A1   A2   A3

A4   A5   A6

Fig. J.1   MMCC™ FRR test samples A1–A6.

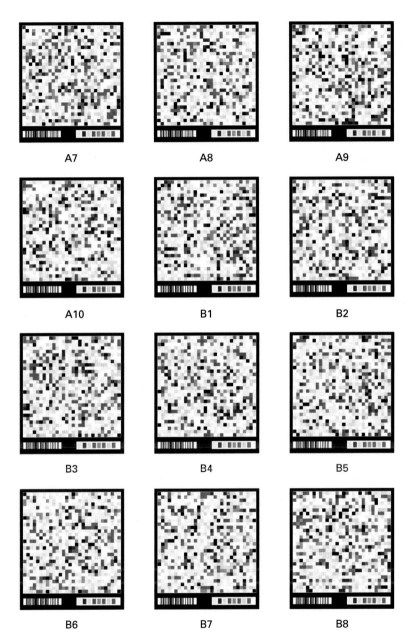

Fig. J.2 MMCC™ FRR test samples A7–B8.

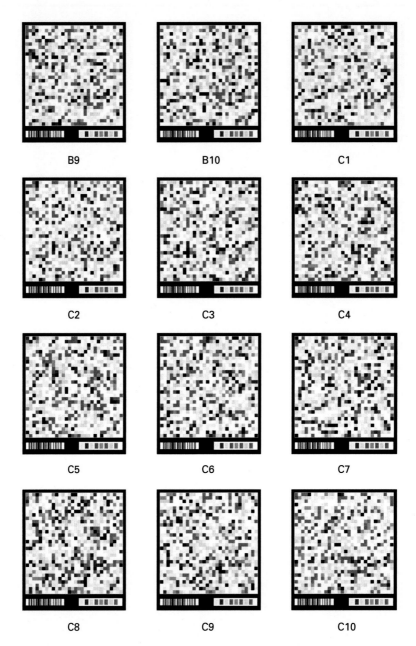

B9   B10   C1

C2   C3   C4

C5   C6   C7

C8   C9   C10

Fig. J.3   MMCC™ FRR test samples B9–C10.

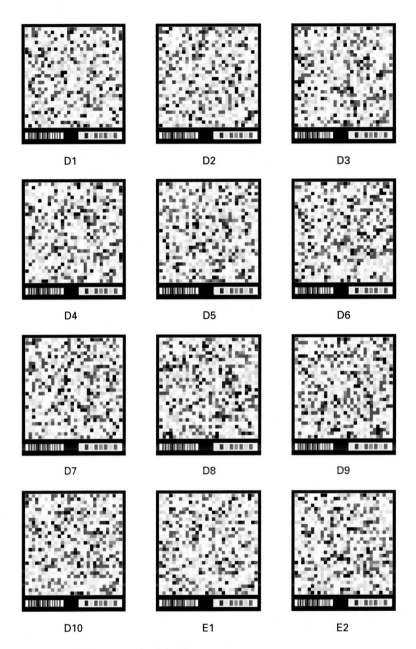

Fig. J.4 MMCC™ FRR test samples D1–E2.

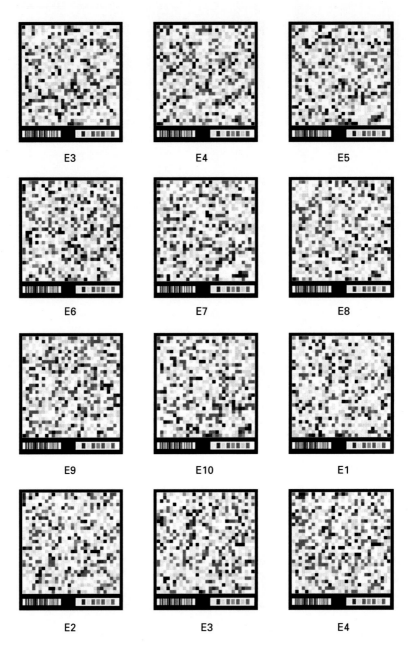

E3   E4   E5

E6   E7   E8

E9   E10   E1

E2   E3   E4

Fig. J.5   MMCC™ FRR test samples E3–F4.

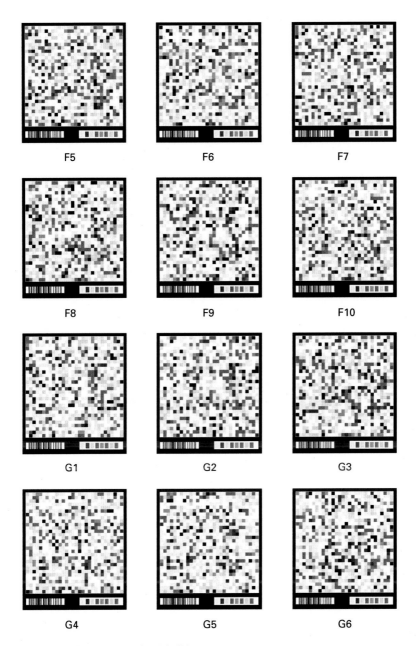

Fig. J.6 MMCC™ FRR test samples F5–G6.

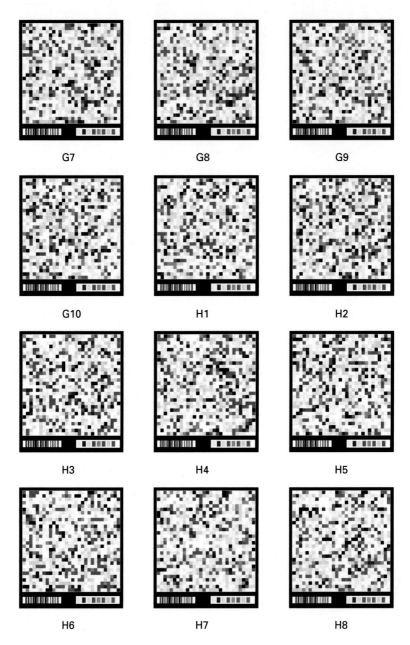

Fig. J.7   MMCC™ FRR test samples G7–H8.

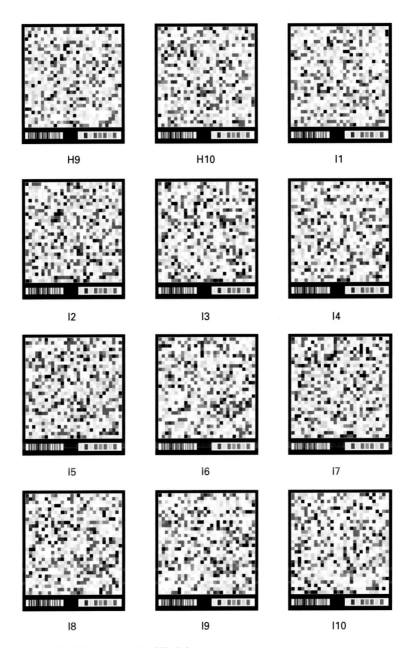

Fig. J.8   MMCC™ FRR test samples H9–I10.

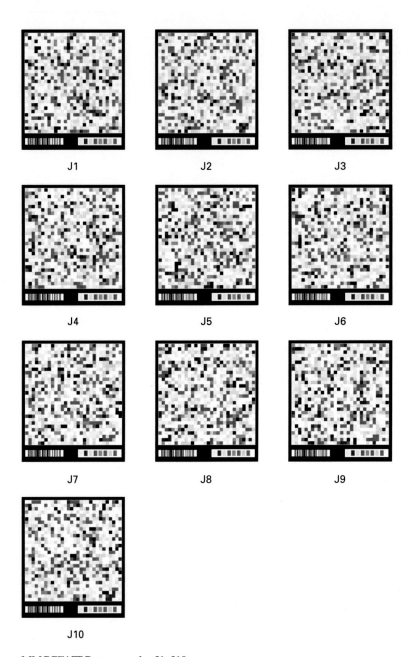

J1　　　　　　　J2　　　　　　　J3

J4　　　　　　　J5　　　　　　　J6

J7　　　　　　　J8　　　　　　　J9

J10

Fig. J.9　　MMCC™ FRR test samples J1–J10.

Appendix K　Subsampling method

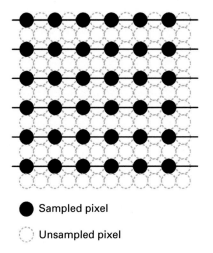

● Sampled pixel

○ Unsampled pixel

Fig. K.1　Subsampling method used in the first and second experiments for evaluation of the MMCC™ barcode.

References

[1] Bushnell, R. D. and Meyers, R. B. *Getting Started with Bar Codes: A Systematic Guide*, 5th edn (Pennsylvania, USA: QuadII, Inc., 1999).

[2] Japanese Standards Association. Glossary of terms relating to data carriers, Japanese Industrial Standards X 0500. 2002-03-20.

[3] Adams, R. E. BarCode1 [online] 2007 [accessed 22 Jan 2008]. Available from http://www.adams1.com/pub/russadam/barcode1.html.

[4] LaMoreaux, R. D. *Barcodes and Other Automatic Identification Systems* (Surrey: Pira International, 1995).

[5] The Health Industry Business Communications Council®. The Health Industry Number System (HIN®) [online] 2008 [accessed 7 Jun 2008]. Available from http://www.hibcc.org/hinsystem.htm.

[6] GS1 US Inc. *BarCodes and eCom*™ [online] 2008 [accessed 5 Aug 2008]. Available from http://barcodes.gs1us.org/dnn_bcec/Default.aspx?tabid=168.

[7] Rekimoto, J. and Ayatsuka, Y. CyberCode: designing augmented reality environments with visual tags, in *Proc. Conf. on Designing Augmented Reality Environments (DARE'00)*, 2000, 1–10.

[8] Bushnell, R. D. and Meyers, R. B. *Sourcebook of Automatic Identification and Data Collection* (New York: Van Nostrand Reinhold, 1990).

[9] Woodland, N. J. and Silver, B. Classifying Apparatus and Method, United States Patent 2 612 994. 1952-10-07.

[10] Hiramoto, J. *Information on Barcode and 2D Symbologies* (Tokyo: Japan Industrial Publishing Co., 2006).

[11] Collins, D. J. *Using Bar Code – Why It's Taking Over* (Duxbury, MA: Data Capture Institute, 1990).

[12] Distribution Systems Research Institute. バーコードのおはなし (Barcode Details) (Tokyo: Japanese Standards Association, 1990).

[13] Japan Automatic Identification Systems Association. これでわかった２次元シンボル -バーコードのすべて- *(Understanding 2D Symbologies: Detailed Information On Barcodes)* (Tokyo: Ohmsha, 2004).

[14] ISO/IEC SC 31 National Body of Japan. New work item proposal Micro QR Code Inclusion to ISO/IEC 18004 QR Code, ISO/IEC SC 31, National Body of Japan. 2003-05-14.

[15] Barnes, D., Bradshaw, J., Day, L., Schott, T. and Wilson, R. Two dimensional bar coding. Purdue University technical paper 621. 1999.

[16] Thoresson, J. PhotoPhone entertainment, in *Proc. Conf. on Human Factors in Computing Systems (CHI'03)*, 2003, 896–897.

[17] Rohs, M. C. Real-world interaction with camera-phones, in *Proc. 2nd Int. Symp. on Ubiquitous Computing Systems (UCS 2004)*, 2004, 74–89.

[18] Tokura, N. Barcode tutorial for beginners. *Barcode*, **1**, 2002 (Tokyo, Japan: Japan Industrial Publishing Co.,).

[19] Nguyen, T. H., Milanesi, C., Liang, A. *et al. Forecast: Camera Phones, Worldwide, 2004–2011* (Connecticut, USA: Gartner, 2007).

[20] López de Ipi, D., Mendonça, P. R. S. and Hopper, A. TRIP: a low-cost vision-based location system for ubiquitous computing, *Personal Ubiquitous Computing London*, **6**:3, 2002, 206–219.

[21] GS1. About GS1 [online] [accessed 7 Jul 2008]. Available from http://www.gs1.org/about/.

[22] Want, R. An introduction to RFID technology, *IEEE Pervasive Computing*, **5**:1, 2006, 25–33.

[23] Mattern, F. Wireless future: ubiquitous computing, in *Proc. Wireless Congress*, Munich, 2004.

[24] Denso Wave Inc. *QR Code.Com* [online] 2003 [accessed 17 Mar 2008]. Available from http://www.denso-wave.com/qrcode/index-e.html.

[25] International Organization for Standardization. Information technology – automatic identification and data capture techniques QR Code 2005 barcode symbology, ISO/IEC 18004. 2006-08-31.

[26] Saito, K. J-SH09 – a camera becomes a bridge between virtual world and real world [online] 2002 [accessed 20 Mar 2008]. Available from ITmedia Inc. at http://plusd.itmedia.co.jp/mobile/0208/01/barcode.html.

[27] Japanese Industrial Standards. Two dimensional symbol – QR Code – Basic specification, JIS X 0510. 2004-11-20.

[28] International Organization for Standardization. Information technology – International symbology specification – Data Matrix. ISO/IEC 16022. 2006-09-15.

[29] AINIX Co. Characteristics of 2D symbologies [online] 2002 [accessed 07 Apr 2008]. Available from http://www.ainix.co.jp/.

[30] Semacode Co. All the technical details you could possibly want, and more [online] 2006 [accessed 8 Apr 2008]. Available from http://semacode.com/.

[31] Veritec Inc. Benefits of Veritec's 2D Codes – VeriCode® and VSCode® [online] 2006 [accessed 12 Apr 2008]. Available from http://www.veritecinc.com/vericode.html.

[32] Squires, S. R. and Levinger, J. K. Efficient finder patterns and methods for application to 2D machine vision problems, United States Patent application 20060269136 A1. 2006-11-30.

[33] Lark Computers. Trillcode [online] 2006 [accessed 21 Aug 2008]. Available from http://www.trillcode.com/.

[34] Madhavapeddy, A., Scott, D., Sharp, R. and Upton, E. Using camera-phone to enhance human–computer interaction, in *Proc. 6th Int. Conf. on Ubiquitous Computing (UbiComp 2004)*, 2004.

[35] Scott, D., Sharp, R., Madhavapeddy, A. and Upton, E. Using visual tags to bypass Bluetooth device discovery, *ACM SIGMOBILE Mobile Computing and Communications Review*, 2005, 41–53.

[36] OP3. *ShotCode* [online] 2008 [accessed 18 Jun 2008]. Available from http://www.shotcode.com/.

[37] Kato, H. and Tan, K. T. Pervasive 2D barcodes for camera phone applications, *IEEE Pervasive Computing*, **6**:4, 2007, 76–85.

[38] Maass, R. and Meyer, A. Device and method for access of content by a barcode, World Intellectual Property Organisation PCT/EP2007/052544. 2007-09-27.

[39] Connvision. Swiss Post uses again BeeTagg for apprentices campaign [online] 2008 [accessed 3 Sep 2008]. Available from http://www.beetagg.com/News/?id=76-BeeTagg-Swiss-Post-Switzerland-Apprentices.

[40] Han, T. D., Cheong, C. H., Lee, N. K. and Shin, E. D. Machine readable code image and method of encoding and decoding the same, United States Patent 7 020 327. 2006-03-28.

[41] ColorZip™ Japan Inc. Technical report: Introduction of ColorCode™ [online] 2006 [accessed 20 Apr 2008]. Available from http://www.colorzip.co.jp/ja/.

[42] Gavin, J. System and method for encoding high density geometric symbol set, United States Patent application 20050285761 A1. 2005-12-29.

[43] Goodloe, A., McDougall, M., Gunter, C. A. and Alur, R. Predictable programs in barcodes, in *Proc. 2002 Int. Conf. on Compilers, Architecture and Synthesis for Embedded Systems*, 2002, 298–303.

[44] Yeh, C. T. and Chen, L. H. A system for a new two-dimensional code: Secure 2D code, *Machine Vision and Applications*, **2**, 1998, 74–82.

[45] Intacta Technologies, Inc. INTACTA CODE™ enhanced security management (ESM). 2002.

[46] Weiser, M. The computer for the 21st century, *Scientific American*, **265**, 1991, 94–104.

[47] Commonwealth of Australia. Australia's regulator for broadcasting, the Internet, radiocommunications and telecommunications: Vision 2020 [online] 2007 [accessed 7 Aug 2008]. Available from http://www.acma.gov.au/WEB/STANDARD/1001/pc=$PC_1$498.

[48] Siio, I., Masui, T. and Fukuchi, K. Real-world interaction using the FieldMouse, in *Proc. 12th Annual ACM Symp. User Interface Software and Technology (UIST'99)*, 1999, 113–119.

[49] Kohtake, N., Rekimoto, J. and Anzai, Y. InfoStick: an interaction device for inter-appliance computing, in *Lecture Notes in Computer Science*, **1707**, 1999, 246–258.

[50] Ayatsuka, Y. and Rekimoto, J. Active CyberCode: a directly controllable 2D code, in *Proc. Human Factors in Computing Systems (CHI'06)*, 2006, 490–495.

[51] Fitzgibbon, A., Pilu, M. and Fisher, R. B. Direct least square fitting of ellipses, *IEEE Trans. Pattern Analysis and Machine Intelligence*, **21**:5, 1999, 476–480.

[52] Shekar, S., and Nair, P. and Helal, A. S. iGrocer: a ubiquitous and pervasive smart grocery shopping system, in *Proc. 2003 ACM Symp. on Applied Computing*, 2003, 645–652.

[53] Chai, D. and Hock, F. Locating and decoding EAN-13 barcodes from images captured by digital cameras, in *Proc. 2005 5th Int. Conf. on Information, Communications and Signal Processing*, 2005, 1595–1599.

[54] Ohbuchi, E., Hanaizumi, H. and Hock, L. A. Barcode readers using the camera device in mobile phones, in *Proc. 2004 Int. Conf. on Cyberworlds*, 2004, 260–265.

[55] The Economist Newspaper. Phones with eyes. [online] 2005 [accessed 15 Sep 2006]. Available from http://economist.com/science/tq/displayStory.cfm?storyid=3714023.

[56] McCune, J. M., Perrig, A. and Reiter, M. K. Seeing is believing: using camera phones for human-verifiable authentication, in *Proc. 2005 IEEE Symp. on Security and Privacy*, 2005, 110–124.

[57] Han, T.-D., Cheong, C., Yoon, H.-M. *et al.* Implementation of new services to support ubiquitous computing for town life, in *Proc. 3rd IEEE Workshop on Software Technologies for Future Embedded and Ubiquitous Systems (SEUS'05)*, 2005, 45–49.

[58] Gonzalez, R. C., Woods, R. E. and Eddins, S. L. *Digital Image Processing Using MATLAB*, 1st edn (Upper Saddle River, NJ: Prentice Hall, 2004).

[59] Onoda, T. and Miwa, K. Hierarchised two-dimensional code, creation method thereof, and read method thereof, World Intellectual Property Organization PCT/JP2006/307091. 2006-04-04.

[60] Morelos-Zaragoza, R. H. *The Art of Error Correcting Coding*, 2nd edn (Chichester: John Wiley & Sons, 2006).

[61] Eto, Y. and Kaneko, T. 誤り訂正と其の応用 *(Error correcting Codes and their Applications)* (Tokyo: Ohmsha, 1996).

[62] Kato, H., Tan, K. T. and Chai, D. Development of a novel finder pattern for effective color 2D barcode detection, in *Proc. Int. Symp. Parallel and Distributed Processing with Applications (ISPA'08)*, 2008, 1006–1013.

[63] Kozaki, Y. and Nishii, Y. *Mechanism of Digital Cameras* (Tokyo: NikkeiBP Soft Press, 2004).

[64] Gonzalez, R. C. and Woods, R. E. *Digital Image Processing*, 2nd edn (Upper Saddle River, NJ: Prentice Hall, 2002).

[65] Konica Minolta Sensing Inc. The essentials of imaging [online] 2008 [accessed 10 Oct 2008]. Available from http://konicaminolta.jp/instruments/colorknowledge/index.html.

[66] Montufar-Chaveznaza, R. Face tracking using a polling strategy, *Int. J. Applied Science, Engineering and Technology*, **3**:3, 2007, 113–117.

[67] Montufar-Chaveznaza, R., Gallardo, F. H. and Hernandez, S. P. Face detection by polling, in *Proc. 2005 IEEE Int. Workshop on Intelligent Signal Processing*, 2005, 292–297.

[68] Yang, Y., Peng, Y. and Liu, Z. A fast algorithm for YCbCr to RGB conversion, *IEEE Trans. Consumer Electronics*, **53**:4, 2007, 1490–1493.

[69] Symes, P. *Digital Video Compression* (New York: McGraw-Hill, 2004).

[70] Briggs, D. Part 9. The dimensions of brightness, saturation and 'colorfulness', in *The Dimensions of Colour* [online] 2008 [accessed 18 Oct 2008]. Available from http://www.huevaluechroma.com/091.php.

[71] Wellner, P. D. Adaptive thresholding for the DigitalDesk. Technical Report EPC-93-110, Rank Xerox Research Centre, Cambridge, UK. 1993.

[72] Heckbert, P. S. Fundamentals of texture mapping and image warping. Master's thesis, University of California at Berkeley. 1989-06-17.

[73] Rekimoto, J., Ayatsuka, Y. and Hernandez, S. P. Augmented reality using the 2D matrix code, in *Proc. Workshop on Interactive Systems and Software*, 1996.

[74] Kato, H. and Tan, K.T. First read rate analysis of 2D-barcodes for camera phone applications as a ubiquitous computing tool, *in Proc. 2007 IEEE Region 10 Conf. TENCON 2007*, 2007, 1–4.

Index

Printed in the United States
by Baker & Taylor Publisher Services